Oral History and Digital Humanities

Voice, Access, and Engagement

Edited by
Douglas A. Boyd and Mary A. Larson

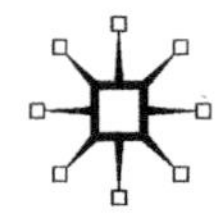

ORAL HISTORY AND DIGITAL HUMANITIES

First published in 2014 by
PALGRAVE MACMILLAN®
in the United States—a division of St. Martin's Press LLC,
175 Fifth Avenue, New York, NY 10010.

Where this book is distributed in the UK, Europe and the rest of the world, this is by Palgrave Macmillan, a division of Macmillan Publishers Limited, registered in England, company number 785998, of Houndmills, Basingstoke, Hampshire RG21 6XS.

Palgrave Macmillan is the global academic imprint of the above companies and has companies and representatives throughout the world.

Palgrave® and Macmillan® are registered trademarks in the United States, the United Kingdom, Europe and other countries.

ISBN: 978–1–137–32200–5 (hc)
ISBN: 978–1–137–32201–2 (pbk)

This book is printed on paper suitable for recycling and made from fully managed and sustained forest sources. Logging, pulping and manufacturing processes are expected to conform to the environmental regulations of the country of origin.

Library of Congress Cataloging-in-Publication Data is available from the Library of Congress.

A catalogue record of the book is available from the British Library.

Design by Newgen Knowledge Works (P) Ltd., Chennai, India.

First edition: December 2014

10 9 8 7 6 5 4 3 2 1

Contents

Figures

Figures

Contributors

Douglas A. Boyd currently serves as the director of the Louie B. Nunn Center for Oral History at the University of Kentucky Libraries and is a recognized leader in the integration of oral history, archives, and digital technologies. Boyd leads the team that envisioned and designed the open-source and free OHMS system that synchronizes text with audio and video online. He recently managed *Oral History in the Digital Age* (http://ohda.matrix.msu.edu) and is the author of the book *Crawfish Bottom: Recovering a Lost Kentucky Community*. He has served on the Executive Council for the Oral History Association (OHA), as the digital initiatives editor for the *Oral History Review*, and recently as chair for the Oral History Section of the Society of American Archivists (SAA). Previously, Boyd managed the Digital Program for the University of Alabama Libraries, served as the director of the Kentucky Oral History Commission, and was senior archivist at the Kentucky Historical Society. Douglas A. Boyd received his MA and PhD in Folklore from Indiana University.

Sherna Berger Gluck writes: "Originally trained in the "Chicago School" of sociology at UCLA, my transition in 1972 from research sociologist to feminist oral historian was natural, although unplanned. It was not only a perfect way for me to forge my social activism with my research interests, but with the founding of the community-based Feminist Research Project, I was able to contribute to the growing number of alternative feminist institutions in Los Angeles. However, to resume earning an income, I re-entered academia in 1974, first introducing a women's oral history course at UCLA, and in 1977 joining the women's studies program at California State University, Long Beach. The following year, in 1978, I founded what became the Oral History Program in the Department of History and remained the director until my retirement in 2005.

Since 1977, I have been writing about women's oral history, and have published numerous oral history related articles and four books, including: *From Parlor to Prison: Five American Suffragists Talk about their Lives* (1976); *Rosie the Riveter Revisited: Women the War and Social Change* (1987); *Women's Words: The Feminist Practice of Oral History* (co-edited with Daphne Patai, 1991); and *An American Feminist in Palestine: The Intifada Years* (1994). In 2000, with Kaye

Briegel, I founded the CSULB Virtual Oral/Aural History Archive (www.csulb.edu/voaha); and for the past eight years, I have been focused on issues related to oral history in the digital age, particularly questions about the promises and perils of online archiving."

A member of the West Chester University Department of History since 1990, **Charles Hardy III** began his work in oral history in the late 1970s. The producer of award-winning radio, video, and web-based documentaries, his sound works include the radio series *I Remember When: Times Gone But Not Forgotten* (1983), *Goin' North: Tales of the Great Migration* (1985), and *The Return of the Shad* (1992); audio art works *Mordecai Mordant's Celebrated Audio Ephemera* (1984) and *This Car to the Ballpark* (1986); and "'I Can Almost See the Lights of Home': A Field Trip to Harlan County, Kentucky" (*Journal of MultiMedia History*, vol. 2, 1999), co-authored with Alessandro Portelli. His video documentary credits include *The United States History Video Collection* (Schlessinger Video Productions, 1996), and *All Aboard For Philadelphia* (Historical Society of Pennsylvania, 1989). His print publications on oral history include "Painting in Sound: Aural History and Audio Art," in *Oral History: The Challenges of Dialogue* (2009); "Authoring in Sound: Aural History, Radio, and the Digital Revolution," in *The Oral History Reader*, 2nd edition (2006); "A People's History of Industrial Philadelphia: Reflections on Community Oral History and the Uses of the Past," *Oral History Review* 33:1 (Winter/Spring 2006); "Oral History in Sound and Moving Image Documentaries" (with Pamela Dean) *Handbook of Oral History* (2006); and "Prodigal Sons, Trap Doors, and Painted Women: Reflections on Life Stories, Urban Legends, and Aural History," *Oral History* 29:1 (Spring 2001). Since 2003, he has served as supervising historian for *ExplorePAhistory*, a collaborative state history website that builds historical content and lesson plans around state historical markers. President of the Oral History Association in 2008–2009, he also served on the Advisory Board of *Oral History in the Digital Age* from 2009 to 2012.

Tom Ikeda is a third-generation Japanese American whose grandparents came to Seattle from Japan in the early 1900s. He is the founding executive director of Densho: The Japanese American Legacy Project, which was started in 1996. Densho uses digital technology to preserve and make accessible primary source materials on the World War II incarceration of Japanese Americans. Densho presents these materials and related resources for their historic value and as a means of exploring issues of democracy, intolerance, wartime hysteria, civil rights and the responsibilities of citizenship in our increasingly global society.

Over the last 18 years, Ikeda has conducted in excess of 200 video-recorded, oral history interviews. He has created online and classroom curriculum from these materials and helped design Densho's award winning website. Prior to Densho, Ikeda was a general manager at Microsoft in the Multimedia Publishing Group.

Ikeda has received numerous awards for his contributions in the humanities, education, and the non-profit sector, including the Humanities Washington Award for outstanding achievement in the public humanities, the Microsoft Technology for Good Award, the National JACL Japanese American of the Biennium award for Education, the Microsoft Alumni Foundation Integral Fellows award, the Japanese Foreign Minister's Award, and the Society of American Archivists Hamer Kegan Award for increasing public awareness through the use of archival materials.

Mary Larson is the associate dean for Special Collections at the Oklahoma State University (OSU) Library, where she joined the faculty in July 2009 as director of the Oklahoma Oral History Research Program. She has been conducting oral histories for 25 years, having previously worked with the programs at two other land-grant institutions—the University of Alaska Fairbanks and the University of Nevada, Reno. Larson is a past president of the Oral History Association and has also served on that organization's Council as well as on the board of the Southwest Oral History Association and as an editor of the H-Oralhist listserv. Her research has centered on a range of topics, including evolving technology and ethical concerns in oral history, the utilization of new media formats for accessibility, and community and rural women's history. Larson holds an MA and PhD in anthropology from Brown University, where her focus was ethnohistory.

Elinor Mazé served for many years as senior editor of the Baylor University Institute for Oral History, where she managed the archiving and presentation of the institute's collection of oral memoirs. She has also worked as a reference librarian, an English language teacher, and a systems analyst. She is the author of "The Uneasy Page: Transcribing and Editing Oral History" in the *Handbook for Oral History* (ed. Rebecca Sharpless, Lois Myers, and Thomas Charlton (Lanham, MD: AltaMira Press, 2006), 237–271), and has been a frequent participant in the annual meetings of the Oral History Association.

Marjorie L. McLellan earned an MA in American Folk Cultures from Cooperstown Graduate Programs and a PhD in American Studies from the University of Minnesota. Formerly an associate professor of history at Miami University, she directed the public history program in the Department of History at Wright State University where she is now an associate professor of Urban Affairs and Geography. A recipient of grants from state humanity councils, the National Endowment for the Humanities, the Institute for Museums and Library Services, the American Association for State Colleges and Universities, and the US Department of Education, she recently completed a two-year term as chair of the Ohio Humanities board. McLellan worked for the Wisconsin Historical Society, and she has worked with numerous historical organizations as instructor, board member, and consultant. McLellan has worked extensively with the public schools including co-directing two Teaching American History projects, collaborating in the development of a large, urban National History Day program, and now promoting civic learning in the Dayton Public Schools. She is the author of *Six Generations Here...A Farm Family Remembers*, *Hunting for Everyday History: A Field Guide for Teachers*, and articles about oral, public, and digital history. The co-founder and advisor of a Youth and Community Engagement minor, McLellan currently serves as interim director of service-learning and civic engagement at Wright State University.

Dean Rehberger is the director of MATRIX: the Center for Humane Art, Letters, and Social Science Online at Michigan State University. Matrix is devoted to the application of new technologies for teaching, research, and outreach. His primary areas of research include: big data and the humanities; oral history online; information design and architecture; digital libraries, museums and archives; and online publishing and learning environments. He oversees the development of a number of open-access projects in the humanities including Slave Biographies (http://slavebiographies.org/), Oral History in the Digital Age (http://ohda.matrix.msu.edu), the MSU Vietnam Group Archive (http://projects.matrix.msu.edu/vietnam/), American Black Journal (http://abj.matrix.msu.edu/), Quilt Index (http://www.quiltindex.org), Overcoming Apartheid (http://overcomingapartheid.msu.edu/), African Oral Narratives (http://www.aodl.org/oralnarratives/), and many other projects found at (http://matrix.msu.edu).

William Schneider writes of himself: "Steeped in what I thought was a well-rounded education in cultural anthropology, I first arrived in Alaska in 1972 and embarked on a new education, mating academic training with an understanding of the lives of rural Alaska Natives. The education curve was steep. I had to learn how to listen, how to negotiate ethical concerns of the community, and basic skills of wood cutting to feed a wood stove that always demanded more than I could provide. I was told that I was welcome as long as I didn't put people under a microscope. The early work that came from that period of my life was a reconstruction of local history based on the stories people chose to tell of how they came and settled in the village of Beaver, Alaska. This was a good training ground for my later work at the University of Alaska Fairbanks where I was curator of Oral History for 30 years.

During my time at the university we pioneered the transition to digital technology for accessing and preserving oral history. In those years I was inspired by the work of folklorists whose explorations into the how, why, when, and where we tell stories raised questions about what we were actually preserving in our oral history. Digital development gave us new ways to organize context and background information to address folkloric concerns and provide a fuller record. But none of this new exploration and development would have been possible without the creative energy of the people who worked with me: Mary Larson, Karen Brewster, and David Krupa. They brought the potentials to fruition."

Stephen Sloan is the director of the Institute for Oral History and an Associate Professor of History at Baylor University. He specializes in oral history, the United States post-1945, public history, and environmental history. Sloan received his PhD from the public history program at Arizona State University. He is a past president of the Oral History Association and has authored several pieces for the *Oral History Review*. He is also the co-editor of *Listening on the Edge: Oral History in the Aftermath of Crisis* published by Oxford Press in 2014. His work with oral history has been funded by grants at the local, state, and national levels, and he has presented his research at many state and national meetings and abroad at conferences in Liverpool, Prague, Guadalajara, Naples, Istanbul, Buenos Aires, and Barcelona.

Gerald Zahavi is professor of History and director of the Documentary Studies Program at the University at Albany, State University of New York, where he has been since 1985. He is the author of *Workers, Managers, and Welfare Capitalism: The Shoemakers and Tanners of Endicott Johnson, 1890–1950* (University of Illinois Press, 1988) and is nearing the completion of another book on the local and regional history of American communism titled *Embers on the Land* (to be published by the University of North Carolina Press). He is also the author of a number of articles on the history of labor and radicalism. In addition to his work on monographs and articles, Zahavi has also been engaged in a variety of media projects. He was the producer and audio engineer of a two-CD oral history of the Lamont-Doherty Earth Observatory of Columbia University and has also been heavily involved in several document and media preservation and publication projects, serving as the editor of over half-a-dozen labor and business history-related microform publications and several history websites. As an active radio broadcaster, he co-produces several programs, including *Talking History*, a weekly radio show broadcast by RPI and the University at Albany, SUNY. In 2006, Zahavi helped establish an interdisciplinary Documentary Studies Program at the University at Albany, a program that he has directed since that time. He also helped inaugurate, in 2009, an innovative History and Media M.A. concentration that offered research and production training for history graduate students interested in cutting-edge work in history and hypermedia authoring, photography and photoanalysis, documentary video/filmmaking, oral/video history, and aural history and audio documentary production. Courses created under that concentration are now part of the Department's recently expanded and revamped Public History Program.

Preface

The editors first met in 1998 at the Society for American Archivists (SAA) meeting in Orlando, Florida, and instantly connected over conversations about oral history, access, and digital technologies. Since then, we have repeatedly served on panels and committees together, continuing these conversations. Based on a shared interest in interdisciplinary approaches, our conversations in the past few years began to center on oral history's impact and role in emerging trends pertaining to digital humanities (DH). Following discussions in the DH community, we continually questioned why oral history was not actively part of those dialogues. This book is our way of honoring oral history's early digital revolutionaries and their sense of experimentation and adventure while also firmly placing these activities within the current digital humanities conversation.

Specifically, we would like to thank:

Linda Shopes and Bruce Stave, the editors for the oral history series at Palgrave when we began this journey.

Contributing authors Sherna Berger Gluck, Charles Hardy, Tom Ikeda, Elinor Mazé, Marjorie McLellan, Dean Rehberger, William Schneider, Stephen Sloan, and Gerald Zahavi.

Douglas A. Boyd would like to thank Jennie, Charlotte, Kathleen, and Eleanor Boyd for their love and support. Terry L. Birdwhistell has offered friendship, leadership, and tremendous service to the oral history field. I am grateful to both Terry and Deirdre Scaggs at the University of Kentucky Libraries for their ongoing support of oral history and innovation.

Mary A. Larson wishes to express her undying gratitude to Mark McDonald and Dorothy Larson for their support. She also thanks her early mentor, Bill Schneider, for his good counsel and creative approaches to oral history, and Sheila Johnson and Jennifer Paustenbaugh, for their continued encouragement.

Introduction

Douglas A. Boyd and Mary A. Larson

The human voice, in the act of meaningful communication, consists of carefully crafted and culturally shaped pressure waves traveling through the air in the form of words, woven together in the form of a story. Stories, formed by memory and performed in narrative either resonate and engage, are possibly preserved and imprinted in memory, or they go unremembered and are lost to time. History is made up of the stories of humanity, based on fragments preserved in time. While material culture—architecture, art, broken clay pots—along with the written words—diaries, records, books—leave a tangible, touchable inheritance for those seeking to understand the past, the spoken word performed in the form of stories has traditionally proven more elusive to preserve. Yet, it is the voice, the first-hand accounts, and the privilege and opportunity for scholars to ask direct questions and grapple with spoken answers of the past that the historian seeks.

One of the major issues throughout oral history's practice, and something that has almost served to define its various phases, has been how an oral history has been represented as an entity. This has been a source of conversation and debate from the inception of professional oral history associations, and largely, it has been mediated by technology, so our very understanding of what an oral history actually *is* has been altered through time based on the tools at practitioners' disposals. Historians, folklorists, digital humanists, ethnologists, anthropologists, and archivists of the modern era have utilized an expanding range of technologies to collect preserve, understand, interpret, and retell stories. This book addresses the history of that process within oral history in the United States and examines how it connects with digital humanities scholarship.

As time has passed, innovative and creative technologies have emerged to transform our methods of preserving and presenting stories. Microphones and recording machines—the wax cylinder, the wire recorder, the reel-to-reel, and its portable successor, the cassette recorder—brought the voice back into the documentation of and engagement with history's performed stories. It was technology from which oral history was born. Once limited by memory, paper, and ink, interviewers mechanically recorded stories of the past and promised to give voice to individual actors of history.

As analog recording technologies grew more affordable and accessible, oral history practice rapidly increased. By the 1960s, oral history had clearly emerged as a compelling methodology for documenting and understanding the individual in the study of history. As the methodology grew more popular, important existential questions arose in the professional community, yielding numerous debates and discussions among leaders and practitioners of oral history. One such debate emerged regarding the role of the recordings and the role of transcripts in the practice and in the purpose of oral history. With a largely dominant text-based focus, early practitioners would destroy the recordings once they were transcribed, often binding, shelving, and cataloging the transcript in library model derived from the book. On September 25–28, 1966, oral history's first generation of revolutionaries came together in Lake Arrowhead, California, to discuss a wide range of major issues and challenges. As documented in *Oral History at Arrowhead: The Proceedings of the First National Colloquium on Oral History*, Knox Mellon questioned the practice of destroying the tapes. Elizabeth Dixon, head of the Oral History Program at UCLA posed a practical response:

> One thing is economy. You keep buying tape, and we're back to the budget again! We can't afford it. Another thing, as Dr. Brooks has said, is that many people would not give you such candid tapes, if they thought you were going to keep them forever because they may not like the way they sound on tape. Conversation is not grammatical. Many times they make errors of statement that can be corrected in a transcript but would have to be spliced out in a tape.[1]

UCLA folklorist Wayland Hand countered that, "Folklorists have made a fetish of the received word exactly as it comes from the lips of the informant. Any tampering with it is condemned... We do generally keep the tapes and we don't tamper with them."[2] Louis Starr, director of the Columbia University Oral History Research Office, on the other hand, believed that:

> It's foolish to imagine that it's going to be worth saving fifty tapes of Francis Perkins. When you want to see exactly how she said it on page two thousand and sixty-three, you're not going to be able to find that place on the tape for a whole half hour or so. By the time you have, you'll decide it wasn't worth the trouble.[3]

Later on in the sessions, Louis Shores, dean of the Library School at Florida State University, pondered the question from the library/archival perspective:

> The introduction of the tape recorder in 1948 did, in my opinion, offer an approach to the record through another medium than writing. Although we have in many cases eliminated this new format for history by insisting on erasing the tape, once it has been transcribed, or by prohibiting the loan of the tape, or by assuming that the transcription is really the primary source...is it not possible that the distilling of the tape into a typescript has, even with the highest integrity and devotion, resulted in the modification of the primary source, the tape? Doesn't strict allegiance to historical bibliography dictate that we acknowledge the typescript to be a secondary rather than a primary source? But above all, should not our oral history custodianship insist upon the preservation of the original tape?[4]

In scholarly hindsight, Columbia's early practice of destroying the original tapes has been, traditionally, vilified. However, although it is not always framed as such, Louis Starr's defense of the practice was primarily framed by the contemporary notion of access and usability. Starr later added in 1977 that, "Tapes, no matter how carefully indexed, are awkward to use," qualifying, "Future generations may prove more aurally oriented."[5] Quite simply, text in the form of the transcript has historically posed the most efficient human interface for long-form oral history interviews. The notion of requiring the analog user/researcher to navigate thousands of pages of typescript and then to expect them to mechanically seek the corresponding moment on a reel-to-reel tape seemed preposterous to most at the time. Although best practice quickly turned away from the destruction of the recordings following transcription (mostly advocated by the archival community), the elements of economics with regard to preservation and the usability and reliability debates surrounding transcription continue today. Our personal observations of oral history's use in the archive over the course of the past two decades support this claim. The transcript, whether in the form of typescript or textual data, is easier and more efficient for a person to navigate, browse, or search. The problems with transcripts remain the fact that they are too expensive to produce on a mass scale, and, quite frankly, they are imperfect representations of the recorded interview.

As oral historians accelerated collection efforts, archivists grappled with the practicalities of preserving and providing access to oral history. Audio and video have traditionally been very difficult and expensive to curate in an archival setting. The formats were fragile and proprietary and, as technologies advanced, the threats of compatibility and obsolescence grew. The greatest challenge oral history faced in the analog archive, however, was the threat of obscurity. Unlike other text or graphic-based archival formats, oral history archives struggled to overcome the logistics of discovery and access, a set of mysteries created when archiving audio and moving images in the absence of text. Without the transcript, the archive might have no more information about an oral history interview on its shelves beyond a name, a date, and the association with a particular project. Archives simply do not have the time or resources to actually listen to each and every moment in each and every interview in order to provide accurate and useful descriptions of the contents to their researchers. As projects, collections, and archives grew in numbers, the typical analog-based oral history archive grew more inaccessible. Because of difficulties with usability and discoverability, archival reuse of oral history projects was limited to research projects yielding academic books and articles, with a few intermittent radio and television documentaries. A typical oral history interview contains a massive amount of information—questions, answers, description, reflection, dialogue, laughter, silences, language, culture, worldview—yet, from the researcher's perspective, oral history's greatest value is found in the moments. Without text in the form of descriptive metadata, transcripts, or indexes, these recorded moments, dormant in the analog archive, unpredictable and inaccessible, were ignored.

Over the last two decades, much has changed in the world of oral history. Through technological advances, the Internet has become a practical way of making recorded sound and video available, opening up a wide range of possibilities for the presentation of material. The Internet has, quite frankly, blown the hinges from doors of the archives, and *access* has come to have a completely different meaning. Oral history research no longer necessarily entails traveling from one university or museum to another, reading typewritten transcripts or listening to second-generation analog cassettes designated as user copies. Scholars and the general public alike can go to the Internet for information on available recordings, and even if an institution's audio files or transcripts are not online, chances are good that a detailed finding aid is, and a call to that repository can result in a patron receiving a digital copy of an interview. Networked information and linked data empower users and researchers to connect oral history resources to other relevant resources. All of this results in expectations for access that are vastly different than they were a mere 20 years ago.

We now have incredible technologies that can disseminate oral history to a global audience almost instantaneously. Archives that once boasted hundreds of annual users of their collections now regularly track thousands of uses of their oral history interviews all over the world. Media outlets such as YouTube or SoundCloud offer near instant and free distribution of audio and video oral histories, while digital repository and content management systems like CONTENTdm or Omeka, or even Drupal or Wordpress, provide powerful infrastructure for housing oral histories in a digital archive or library. Systems such as OHMS (Oral History Metadata Synchronizer) now provide free opportunities to enhance access to oral histories online, connecting a textual search of a transcript or an index to the correlating moment in the online audio or video interview. Mobile applications like Curatescape are part of the mix, too, as they can now provide a framework for mobile users to hear oral history, not only in a moment, but in specific locales as well. Digital technologies now offer enormous opportunities for collecting, curating, and disseminating interviews and projects. While they may have solved certain issues of access, preservation, contextualization, and presentation, however, new technologies have also posed concomitant potential threats, including increased vulnerability of narrators, infrastructure obsolescence, and a host of other ethical issues, particularly with heritage collections.

Much has been written about the "digital revolution" and its potential for "transformation" of oral history.[6] The purpose of this book is to allow key innovators to reflect on the methodological and theoretical developments that have occurred in the practice of oral history in the United States since digital audio and video became practical working formats and on how those developments tie in to conversations in digital humanities. So much has changed in how practitioners approach every aspect of an oral history project, from the very beginning stages to the end products. Numerous individuals participated in the emergence of oral history's slice of the digital revolution to varying degrees. Some merely upgraded their recording technologies, while others envisioned new, creative ways for fundamentally engaging with oral history.

Digital technologies posed numerous opportunities to explore new models for automating access and providing contextual frameworks to encourage more meaningful interactions with researchers as well as with community members represented by a particular oral history project. For the purposes of this book, we have identified individuals who have been intimately involved in different stages of oral history's digital shift in the United States. Most of them were not trying to be revolutionary, nor were they going with the methodological flow. Instead, they were concerned by both practical challenges and theoretical problems with regard to the discovery, access, and usability of oral history that the previous analog framework imposed upon them.

All of the authors in the following chapters were involved with profoundly important oral history projects and recognized early that digital technologies provided so much more than an improved recording experience. Each one recognized the importance of the voice in oral history, and they were not afraid to explore or experiment with emerging frameworks in order to find more effective and meaningful ways to connect users/researchers to their powerful interviews. For this book, they have contributed chapters that frame the diachronic march of the methodology, focusing not on the current state of the field so much as reflecting on the changes that occurred during the course of particular projects.

Structure of This Book

The volume begins with this introduction, which provides a contextualization for all that follows. In this chapter, readers will find a discussion of the book's genesis together with an introduction to some of the themes that appear throughout. The rest of the text is divided into three parts—two topical sections, Orality/Aurality and Discovery and Discourse, followed by two final essays, written by oral historian Stephen Sloan and digital humanist Dean Rehberger, that are intended less as capstone chapters and more as starting points for future dialogue.

The digital revolution has impacted almost every facet of oral history except for its one primary feature—the fact that an interview is still a dialogue, created through the interaction of (at least) two human beings, one with a story to tell, and one who wants to hear it. This book aims to extend that concept on a number of fronts. First, as editors we wanted to bring a dialogic approach to the chapters themselves. The authors in the Orality/Aurality and Discovery and Discourse sections have contributed their essays to this volume, but they have also participated in follow-up interviews about their work, which were conducted by the editors. The resulting interviews and chapters have been used as the basis for the two essays that appear at the end of the Orality/Aurality and Discovery and Discourse units (written by Boyd and Larson, respectively), and they bring together some of the main trends and themes discussed in the preceding sections. While only excerpts of the actual interviews appear in these capping chapters, the complete versions will be made available on the website that accompanies this publication (http://www.digitaloralhistory.net).

We also see the rich possibilities inherent in using digital technology to involve readers in an ongoing discussion through the companion website, with links to the interviews conducted for the book as well as to relevant external websites, blogs, social media, or other related formats. We hope that these will provide a virtual town hall for the book, with readers submitting feedback and engaging in extended discussions and exchanges of ideas. This book is no longer, then, limited to what is printed in it at the time of publication, but it has the capacity to evolve through online conversations and additions. In this way, the concept of dialogue comes full circle, in that this is ultimately a book about dialogue, with dialogue that aims to engender further dialogue.

The middle portion of the book consists of the two sections on Orality/Aurality and Discovery and Discourse, and the following description outlines what readers can expect to find in that part of the volume:

Orality/Aurality

For all the discussion that has occurred over the years regarding privileging the transcript or the recording, the analog framework made it extremely difficult to use anything but the text. Additionally, the analog realm provided very few distribution opportunities for recorded audio and video oral history interviews, and traditional media outlets like radio and television permitted only a small number of oral history projects to pass through their gates. In his chapter "Oral History and the Digital Revolution: Toward a Post-Documentary Sensibility," published in 2006, Michael Frisch confirmed Louis Starr's conjecture about *aurality* by positing that new digital technologies and tools would shift oral history's focus to "the actual voice (orality, in all its meanings), and embodied voices and contexts in even richer video documentation, returns to the centre of immediacy and focus in oral history."[7] The emergence of digital infrastructure posed tremendous new opportunities to return the voice to users' and researchers' engagement with oral history.

Beginning in 1988, William Schneider at the University of Alaska Fairbanks (UAF) realized that upcoming budget cuts would mean that he was going to have to do more with less in his program, and he was hoping that a computerized delivery system would help make the oral histories in the UAF collection more accessible. He also hoped that an automated approach would deliver the collection to populations traditionally underserved by the university—particularly Alaska Native communities in rural villages—while simultaneously maintaining the oral aspect of the interviews. At the same time, Schneider was working in cross-cultural settings and understood the importance of contextualizing material. As a way of addressing all of these issues with one platform, he started working with a team at UAF to develop Project Jukebox, which ultimately evolved into a series of multimedia, oral history databases that would later be available online. In his essay, Schneider discusses both the plus side of contextualization and accessibility, while also addressing the accompanying challenges posed by ethical issues and infrastructure obsolescence.

Sherna Berger Gluck was primarily concerned with intentionally creating a digital platform that presented oral histories in such a way as to maintain the unmediated orality/aurality of these interviews. In conversations with one of the editors back in the early 1990s, Gluck expressed a desire to make interview audio accessible while at the same time balancing that availability with an ethical approach (something also closely considered by Schneider). The oral history program at California State University, Long Beach unveiled *The CSULB Virtual Oral/Aural History Archive* (VOAHA) in 2003, with the project's name reflecting its primary audio focus but also maintaining an emphasis on presenting contextualized audio. VOAHA provided an early discovery interface for navigating curated excerpts, carefully selected and comprehensively described for the user. Again, issues of technical obsolescence were involved and threatened the existence of the project, and Gluck talks about the long-term support required on the part of institutions as well as the ongoing ethical concerns in an era of decreased privacy and uncertain politics.

Having worked in radio, and drawing on his previous experiences crafting and producing *This Car to the Ballpark* and *Mordecai Mordant*, Charlie Hardy of

West Chester University was aware of the way that audio oral histories, recorded in hi-fidelity, could be interwoven with other material to present powerful stories. Being digitally savvy, he was also one of the first to see the possibilities and opportunities presented by the evolving technologies, in terms of what they could offer for audio productions. Together with Alessandro Portelli, he produced one of the better known aural essays to appear in the *Journal for MultiMedia History*—" 'I Can Almost See the Lights of Home': A Field Trip to Harlan County, Kentucky"—which won the 1999 Oral History Association Award for a Non-Print Media Project. Hardy and Portelli did not set out, originally, to create a model to profoundly impact digital scholarship in the innovative use of oral history. In his interview for this book, Hardy recalls, "For me, distribution was less important than doing something that was a piece of art."[8] What Hardy and Portelli did produce, in concert with the *Journal for MultiMedia History*, was a scholarly work primarily crafted as an audio montage, secondarily supported with reflexive text. Roy Rosenzweig referred to "Lights of Home" as "a spectacular piece of work—a brilliant melding of form and content and probably the best use of the web for scholarship that I have seen."[9] Scholarly journals, at present in 2014, are still trying to emulate "I Can Almost See the Lights of Home" with regard to its effectiveness and grace in incorporating orality/aurality into published, peer-reviewed scholarship.

In the final essay in the section, "'I Just Want to Click on it to Listen,'" Douglas A. Boyd, director of the Louie B. Nunn Center for Oral History at the University of Kentucky Libraries, uses chapters one through three to frame a contemporary reflection on the shifting role and increased prominence of the audio and video recordings in the archival user experience. Boyd stresses the need for developing new, sustainable, and user-driven archival models and tools, and the importance of shifting paradigms with regard to orality/aurality in the discovery of, access to, and engagement with online oral histories.

Discovery and Discourse

The second section of the book reflects primarily on the concepts of discovery and discourse, encompassing everything from metadata and accessibility theory to the public, pedagogical, and academic conversations that are engendered by the availability of historical resources. Threads concerning the democratization of access and the importance of public scholarship are interwoven throughout these essays, as are meditations on the teaching and presentation of oral history and what discovery and discourse contribute to ongoing discussions.

Marjorie McLellan of Wright State University worked with an early project that addressed at least three different issues—documenting local history, engaging college students in that pursuit, and at the same time involving them in larger questions of method and theory. The Miami Valley Cultural Heritage Project (MVCHP) first appeared on the Internet in 1998, when McLellan was still at Miami University. As Irene Reti noted in her 1999 assessment of the project for the *Oral History Review*, this was not just a traditional oral history website (to the extent that any of them were old enough to be *traditional* at that point), but one that provided resources and exhibits and asked meaningful questions about the nature of oral history.[10] As an active member with the Ohio Humanities Council, McLellan also has a very well defined sense of the interface between oral history and the digital humanities, and of how that has evolved over the years. She, too, raises the specter of institutional support for projects.

Gerald Zahavi, at the University at Albany, was looking for ways of opening up the discourse on both history and pedagogy, "to utilize the promise of digital technologies to expand history's boundaries, merge its forms, and promote and legitimate innovations in teaching and research..."[11] In 1996, he founded Talking History (a center for aural productions), and in 1997 he created the *Journal for MultiMedia History*, which utilized the platform of the Internet as a way of making history accessible to a wider audience through the use of media. While it did not revolutionize historical publications at the time as some may have expected, the interactive, multi-format nature of the journal foreshadowed, in many ways, what professional organizations such as the Oral History Association are now, still, hoping to accomplish with peer-reviewed journals such as the *Oral History Review*.

A consideration of how digital technology has changed the basic capabilities for recording and presenting oral history would not be complete without a discussion of the issues involved with large-scale video projects. Tom Ikeda of Densho has been a pioneer in tackling the problems inherent with presenting and preserving materials for this type of undertaking. Because of the sheer amount of storage space required, as well as issues of technology and standardization, not many groups have been able to use this approach. (The Shoah Foundation is another exception, of course, but their unprecedented level of support and backing renders them an impractical example for most organizations.) An interesting twist is that Ikeda, Geoff Froh, and the Densho staff have created a truly virtual archive, since there is no physical location for researchers to visit. All access is through the Densho website itself, which is distinctly different from most oral history programs at present. Also of particular note is the way that they have diligently addressed ethical issues in the building of this archive.

If projects are to be made available in digital formats, one of the keys to discoverability is accurate and appropriate metadata. As part of her work at the Baylor Institute for Oral History, Elinor Mazé has spent more time than most oral historians have considering the implications of how programs describe their interviews. Drawing on her experience at Baylor with CONTENTdm, she discusses her institution's evolving use of metadata within their content management system, and draws attention to aspects of particular importance, as oral historians seek to make material accessible in meaningful ways to a wide range of potential audiences. Not all metadata is created equal, and she discusses the inherent politics of access and discoverability.

In her chapter, "We All Begin with a Story," Mary A. Larson draws on previous chapters—as well as on her extensive experiences working at the University of Alaska Fairbanks, the University of Nevada Oral History Program, and now at Oklahoma State University—to frame a discussion of digital technologies, access and discovery with regard to digital oral history projects, and the impacts on resulting community, scholarly, and public discourse and engagement.

Into the second decade of digital oral history, we have moved beyond mere recording and digitization. The digital revolution has begun to change how the participants in the oral history process conceptualize projects, how they deal with ethical issues, how they process and preserve their materials, how they think about sound and video, how materials are made accessible, and how they "share authority." In fact, new technologies have made oral history more practical and accessible as a methodology (although admittedly at a price), and they have altered what practitioners expect in terms of audio and video quality, products, and the ways in which oral histories are contextualized and analyzed. They have also placed oral history quietly in the middle of the conversation on the digital humanities.

Oral History and Digital Humanities

Based on observations and conversations with colleagues, it seems fairly clear that the number of oral historians who self-identify as digital humanists is relatively limited, which seems a curious thing. To be sure, it could be due to the fact that not everyone using oral history as a methodology is in a humanities field or is using digital technology, and many historians are themselves just dipping their toes into the digital humanities pool. To those who have long had a foot in both worlds, however, the connections are clear and abundant. In fact, three of the tenets oral historians hold most dear—collaboration, a democratic impulse, and public scholarship—are also three of the leading concerns often cited by digital humanists. Add to this the interdisciplinary (or multidisciplinary) nature of both methodologies, together with the importance of contextualization/curation, and one finds that the two camps have more in common than they would have to separate them.

An emphasis on collaboration is one area where oral history and digital humanities are obviously shifting the traditional model in the academy. In many humanities fields, history included, scholarly production in the past has tended to be generated primarily by single scholars, working alone. While there may be collegiality and genuine friendships, the humanities have not been known for group projects. As Bethany Nowviskie has stated, however, "Collaborative work is a major hallmark of digital humanities practice..."[12] and much of that is done by teams of researchers crossing disciplines, so this is a very different paradigm. *The Digital Humanities Manifesto 2.0* states it even more succinctly: "Digital Humanities = **Co-creation**."[13]

Collaboration in oral history may also be interdisciplinary within institutional settings, as with the *Oral History in the Digital Age* project,[14] but it is even more likely to entail close cooperation between oral historians and individuals or communities, predicated on the concept of shared authority. A perfect example of this type of collaboration is William Schneider's work with Project Jukebox. Even in the early days of digital oral history projects, Schneider was working very closely with local groups on the design and implementation of platforms, and he discusses that process in his essay and follow-up interview. Sherna Gluck was also actively involved in this type of community discussion at the point of making materials available online, which she describes particularly in relation to the Japanese American fishing community at Terminal Island. The importance of recognizing this co-authorial role in interviewing projects is also reflected in the (United States) Oral History Association's *Principles and Best Practices* document, where responsibilities to communities are spelled out under item number eight in the "Post Interview" section.[15]

The movement toward cooperative engagement is probably also closely tied to an egalitarian urge on the part of oral historians and digital humanists, and that will be discussed in more detail momentarily. Whatever the reason, it is clear that collaborative work is important to a large number of practitioners in both groups, as it appears everywhere, from the free sharing of interviews on websites or ideas in THATCamps, to the development of open-source code and platforms such as OHMS (Oral History Metadata Synchronizer) or QGIS (Quantum GIS).[16] In her chapter, Marjorie McLellan speaks very specifically about how collaboration has played out for her, both in a pedagogical setting and in the context of cooperative work with IT specialists at the universities where she has worked.

There is a democratic spirit that is common to oral history and digital humanities that engenders a sense that the materials created, shared, generated, or parsed belong to everyone—not just to the educated or the well-to-do, but to those outside the university walls as well as those within. It is reflected in a common sentiment among oral historians concerning the importance of "history from the bottom up," and it is also mirrored in the way that many digital humanities gatherings are conducted.[17] Across a range of DH events, participants are used to hearing one standard instruction at the beginning: You leave your academic rank at the door. (For an especially clear statement, see the basic THATCamp guidelines for unconferences.[18]) These activities are nonhierarchical, and it does not matter whether those attending are associate professors, IT specialists, university administrators, community members, adjuncts, or grad students; no one gets to pull rank. The digital humanities roots of this democratic impulse are noted in the *Digital Humanities Manifesto 2.0*:

> Digital Humanities have **a utopian core** shaped by its genealogical descent from the counterculture-cyberculture intertwinglings of the 60s and 70s. This is why it affirms the value of the open, the infinite, the expansive, the university/museum/archive/library *without walls*, the democratization of culture and scholarship, even as it affirms the value of large-scale statistically grounded methods (such as cultural analytics) that collapse the boundaries between the humanities and the social and natural sciences.[19]

Conference/unconference organizers are very intentional about conveying the sense that not only is everyone to be treated as equal, but everyone is expected to participate.

For oral historians, this nonhierarchical model of engagement parallels in many ways the strongly held belief in shared authority. It also reflects that the democratic urge is not simply to include people in the dissemination of knowledge at the end of a project, but to include as many diverse groups as possible in the generation of the information in the first place. The democracy extends not just to the concept of equitable access to materials, but to equal participation in their creation. In oral history, this plays out through efforts to record formerly under-documented populations, and Sherna Berger Gluck's essay addresses the important role digital technology played in ensuring that the Feminist Oral History Research Project could "make women's voices heard—not just their words on the printed page, but their unmediated *voices*." Tom Ikeda's chapter also discusses this desire to record less-chronicled populations—in the case of the Densho archive, Japanese Americans who had been incarcerated during World War II—but Densho also wanted to ensure that there was diversity *within* this relatively undocumented group. In developing the (United States) Oral History Association's recent strategic plan, their Council also wanted to be sure that this point be made. The fact that "diverse individuals and organizations" are referenced prominently in the organization's two-sentence mission statement, then, is not simply a coincidence, but rather an indication of the importance of this tenet to what oral history stands for at its core.[20]

While most oral historians consider it essential to give back to a community one has worked with, a more general dissemination of information is also important to both groups, with public scholarship being highly valued by oral historians and digital humanists alike. As Lisa Spiro has observed, "For the digital humanities, information is not a commodity to be controlled, but a social good to be shared and reused,"[21] and this is something that is generally true for oral history, as well. Beyond their shared emphasis on collaboration and a democratic impulse, another reason for the importance of public scholarship may be the fact that both groups have a significant core of practitioners operating in nonacademic settings. Oral historians and digital humanists find themselves employed in an array of positions with governmental agencies, historical societies, arts groups, museums, and other nonprofits, and many of those types of organizations have a stated focus on providing information to the general public.

Even within academic circles, though, there is a growing emphasis on public scholarship of all stripes, and this is something that both Charlie Hardy and Gerry Zahavi realized earlier than most academics. Their essays both touch on a wide array of themes, but one of the things they have in common is that the work they have generated over the years makes authentic oral history materials available in ways that encourage a public conversation on historical topics.

While there are many common threads between oral history and the digital humanities, one last one merits attention, and that is curation—which also appears in many oral history conversations as *context*, with overtones of archival responsibility. They are really different ways of expressing the same thing, and we find the definition given in the *Digital Humanities Manifesto 2.0* very familiar:

> **Curation**
>
> Digital Humanists recognize **curation** as a central feature of the future of the Humanities disciplines... Curation means **making arguments through objects as well as words, images, and sounds**. It implies a spatialization of the sort of critical and narrative tasks that, while not unfamiliar to historians, are fundamentally different when carried out in space—physical, virtual, or both—rather than in language alone. It means becoming engaged in collecting, assembling, sifting, structuring, and interpreting corpora. All of which is to say that we consider curation on par with traditional narrative scholarship.
>
> ...
>
> Curation also implies **custodial responsibilities** with respect to the remains of the past as well as interpretive, *meaning-making responsibilities* with respect to the present and future.[22]

The theme of contextualization, particularly as tied to archival responsibility, arises throughout this book in various ways, in discussions of fulfilling a duty to a community or in reflections on preserving meaning in oral history. But it is not just the authors in this publication that recognize the importance of context, as it is cited specifically in the Oral History Association's *Principles and Best Practices* document, which states:

> Information deemed relevant for the interpretation of the oral history by future users, such as photographs, documents, or other records should be collected, and archivists should make clear to users the availability and connection of these materials to the recorded interview.[23]

The reasons for the critical nature of this archival role are demonstrated in the chapters that follow.

Aside from issues pertaining specifically to orality/aurality, discovery and discourse, or themes shared with the digital humanities, there are other threads that are common to many of the essays in this book, crosscutting these largely artificial organizational divisions. Among these are ethics, responsibility to individuals and communities, and issues of audience. Perhaps more obviously, there are also technological concerns, ranging from the two-edged sword of collaboration and tension with IT departments, pushing the technological envelope, and then, on the other side, technical obsolescence. As can be seen in the essays from Schneider, Gluck, and Zahavi, particularly, the latter two elements are often closely tied, with those out in front taking the majority of the flak for the rest of us. Readers should look for these topics to appear and reappear throughout the book.

Beyond the monograph, the success of contemporary oral history projects is now regularly being measured by metrics pertaining to accessibility, discovery, engagement, usability, reuse, and a project's impact on both community and on scholarship. The once modular and disconnected roles of the interviewer, the scholar, the archivist, and the technologist have converged in exciting ways. Archives are moving away from the role of gatekeepers while shifts in open-access publishing and digital scholarship are dramatically shaping new directions of scholarly and community discourse. Oral history projects today are *digital* projects, built on a tradition established by a generation of creative innovators and explorers.

The essays in this book are meant to focus not on case studies of the oral history projects themselves, but on the reasons those projects were developed in the first place, how the researchers solved the posited problems, and how the solutions evolved over time with advancing technologies. All of these practitioners are still heavily invested in oral history and are aware of the latest developments, so in interviews with them, we have asked them to discuss how the problems that started them on their digital paths are currently being dealt with and what they see for the future of oral history. This collection of stories from the digital revolution presents a glimpse into oral history's innovative role in shaping and forming digital public history and scholarship, community engagement, archival discovery and accessibility, and, ultimately, transforming the roles of individual voices and stories in the shaping and construction of our cultural and historical understanding of the past.

Notes

1. Elizabeth I. Dixon and James V. Mink, *Oral History at Arrowhead, Proceedings of the National Colloquium on Oral History* (1st, Lake Arrowhead, CA, September 25–28, 1966), Oral History Association (1966), 22–23.

2. Ibid., 23.
3. Ibid.
4. Ibid., 39.
5. Louis Starr, "Oral History," in *Encyclopedia of Library and Information Sciences*, ed. Allen Kent et al., vol. 20 (New York: Marcel Dekker, 1977), 443.
6. Alistair Thomson, "Four Paradigm Transformations in Oral History," *Oral History Review*, 34(1) (2007): 68; Doug Boyd, "Achieving the Promise of Oral History in a Digital Age," in *The Oxford Handbook of Oral History*, ed. Donald Ritchie (Oxford: Oxford University Press, 2011), 294–295; Steve High and David Sworn, "After the Interview: The Interpretive Challenges of Oral History Video Indexing," *Digital Studies/Le champ Numérique*, 1(2) (2009), http://www.digitalstudies.org/ojs/index.php/digital_studies/article/view/173/215, accessed June 23, 2014; Michael Frisch, "Oral History and the Digital Revolution: Toward a Post-Documentary Sensibility," in *The Oral History Reader*, ed. Robert B. Perks and Alistair Thomson (London: Routledge, 2nd edition, 2006), 102–114.
7. Frisch, "Oral History and the Digital Revolution," 104.
8. Charles Hardy, Interviewed by Doug Boyd, April 30, 2014. Available online at http://www.digitaloralhistory.net.
9. Roy Rosenzweig, Email to Charles Hardy, April 4, 2001; Gerald Zahavi and Susan L. McCormick, "Digital Scholarship, Peer Review, and Hiring, Promotion and Tenure: A Case Study of the Journal for Multimedia History," in *Digital Scholarship in the Tenure, Promotion, and Review Process*, ed. Deborah Lines Anderson (Armonk, NY: M.E. Sharpe, 2004), 114.
10. Irene Reti, "Oral History on the Web," *Oral History Review*, 26(2) (1999): 147–149.
11. From the "Founding Statement" of the *Journal for MultiMedia History*, http://www.albany.edu/jmmh/, accessed June 27, 2011.
12. Bethany Nowviskie blog, "Where Credit is Due," May 31, 2011, http://nowviskie.org/2011/where-credit-is-due/. Additionally, see: Patrick Svensson, "The Landscape of Digital Humanities," *Digital Humanities Quarterly*, 4(1) (2010), accessed May 3, 2014, http://digitalhumanities.org/dhq/vol/4/1/000080/000080.html; Tom Scheinfeldt, "Stuff Digital Humanists Like: Defining Digital Humanities by its Values," *Found History*, December 2, 2012, http://www.foundhistory.org/2010/12/02/stuff-digital-humanists-like/.
13. *The Digital Humanities Manifesto 2.0*, accessed May 3, 2014, http://manifesto.humanities.ucla.edu/2009/05/29/the-digital-humanities-manifesto-20/, emphasis original.
14. *Oral History in the Digital Age*, http://ohda.matrix.msu.edu/.
15. *Principles and Best Practices*, Oral History Association, adopted October 2009, http://www.oralhistory.org/about/principles-and-practices/.
16. OHMS, http://www.oralhistoryonline.org; QGIS, http://www.qgis.org/en/site/.
17. See, for example, Paul Thompson, *The Voice of the Past: Oral History* (New York: Oxford University Press, 1978); Thomson, "Four Paradigm Transformations in Oral History," 51–53; Voytek Bialkowski, Rebecca Niles, and Alan Galey, "The Digital Humanities Summer Institute and Extra-Institutional Modes of Engagement," *Faculty of Information Quarterly*, 3(3) (2011): 20.
18. "What is a THATCamp?" accessed April 12, 2013, http://thatcamp.org/about/.

19. *The Digital Humanities Manifesto 2.0*, accessed May 3, 2014, http://manifesto.humanities.ucla.edu/2009/05/29/the-digital-humanities-manifesto-20/, emphasis original.
20. *Oral History Association Strategic Plan for 2014 to 2017*, accessed April 4, 2014, http://www.oralhistory.org/wp-content/uploads/2014/04/Oral-History-Association-SHORT-3-12-14.pdf.
21. Lisa Spiro, "This is Why We Fight: Defining the Values of the Digital Humanities," in *Debates in the Digital Humanities*, ed. Matthew K. Gold (Minneapolis: University of Minnesota Press, 2012), e-book.
22. *The Digital Humanities Manifesto 2.0*, accessed May 3, 2014, http://manifesto.humanities.ucla.edu/2009/05/29/the-digital-humanities-manifesto-20/, emphasis original.
23. *Principles and Best Practices*, Oral History Association, adopted October 2009, http://www.oralhistory.org/about/principles-and-practices/, "Post-Interview," section 3.

PART I

Orality/Aurality

CHAPTER 1

Oral History in the Age of Digital Possibilities

William Schneider

Typical archival institutions are delivering oral history collections online using repository systems that fail to accommodate oral history's complex, multidimensional nature.

—Doug Boyd[1]

Background

Thirty years ago, digital technology for oral history was in the "Baby Waiting Room" of most oral history programs, and the Internet wasn't even a twinkle in the eye of the pioneering parents who would make it a universal portal to information. At the University of Alaska Fairbanks (UAF), we stumbled onto digital technology for oral history under the false assumption that it would save us money and personnel in the long run, since retrieval, access, and storage could theoretically be done automatically, without human labor. In 1987, the university was going through one of its economic cutbacks, and the Oral History Program was on the chopping block. A graduate student, Felix Vogt, initiated the research that led to an Apple Library of Tomorrow Grant, and that funding provided the necessary equipment to explore digitization. This was the undertaking that would become Project Jukebox. Our first actual developer was Dan Grahek, and his work was premiered at the 1991 meeting of the Oral History Association

(OHA) in Salt Lake City. A dinosaur by today's standards, that standalone station may have been the first time a digital presentation was given at OHA.

Now, with electronic access as the norm for oral history, we look back and ask how this new mode of access has changed the methodology. Of course, this begs two more fundamental questions: what are oral histories? And what are we doing when we "do oral history?" If we begin with these questions, we are less likely to lose track of our central concerns, which are the information and meaning that were originally shared at the time of the recording. A central thesis of this chapter is that we need to be very clear about how we preserve and present oral histories and how that may differ in meaning and intent from what was shared at the time of the recording. These considerations have always been important, but they take on more meaning with digital technology. Through these new platforms, we have made it easier for anyone, at any time, to get access to recordings, and this decreases the likelihood of any in-person dialogue between the interviewer/recorder/collection manager and future listeners. When a researcher has to actually check out an oral history, the collection manager (who, in some cases, did the recording), can provide background information. The loss of this personal contact and thus, the subsequent information transmission, is a possible downside to electronic delivery. On the upside, digital delivery has given us opportunities to add supporting material to help reestablish the setting and background of the account. Because of these possibilities, when the public finds a recording online, they potentially have a great deal of information right there, on their screens, to help them understand what was shared. To do this requires developing a sustainable platform that can deliver these materials over the long term, which could be difficult, considering the ever-changing state of digital technology.

Introduction

While there are always two (and sometimes more) participants in the initial recording of an oral history, I would argue that there are three primary players in the presentation and preservation of a digital oral history once it has been recorded—the oral historian, the collection manager, and the Information Technology (IT) specialist. These three roles may, in some programs, actually be represented by the same person, but there are specific concerns and responsibilities particular to each. I will start with the perspective of the oral historian.

Context plays a big part in the discussion that follows, so I want to provide a little background on the experiences that shape my approach to the issues addressed. Unlike most oral historians, I am trained as a cultural anthropologist, and most of my research is in the field of ethnohistory with Alaskan Native cultural groups. My early work was with life histories based on oral history interviews, a topic that led me to the research of folklorists and their understanding of voice and the challenges of retaining meaning in transcription. Working in cross-cultural contexts has provided me with opportunities to experience the variety of ways in which people use oral narrative to convey meaning to each other, and it has nurtured in me a sensitivity to the nuances that members of the same culture often take for granted, but that can easily be missed by outsiders. My awareness of this has been honed through the anthropological tradition of learning through experience and through the recognition of how easily I could make assumptions about meaning only to realize later that I misunderstood. By appreciating the ambiguity that often surrounds such work, I have gained a strong appreciation for the value of hearing accounts many times over, and in different contexts, in order to understand the meaning—and in some cases, to recognize multiple meanings, depending on context and audience.

While we may think that these considerations are more relevant in cross-cultural settings where we are "out of our element," I think we are challenged to probe meaning in all oral narratives, whether they be familiar or foreign. Digital technology can be a helpful tool to document context, to replicate the nuances of narrator presentations, to provide a comparative record of other tellings, and to provide multi-format supporting information, all in the same searchable and retrievable package. But technology, with its opportunities and constraints, can also take over our attention, and we can get carried away with the possibilities offered and lose track of the speakers and their narratives. That is the principal reason why, when we were starting to develop Project Jukebox, it was important for us to be very clear about what we wanted to preserve and present with digital technology, while at the same time recognizing what we might be losing in the process.

Finding and Retaining Meaning in Oral History

Conventional definitions of oral history focus on two things: a recorded interview and preservation of the recording for future reference and use.[2] Oral histories are composed of interviews, often focusing on personal narratives, and recordings of performances featuring renditions of oral tradition. When I talk about oral history, I try to keep foremost in my mind that any recordings are a shadow of what went on in the original telling. Once a story is recorded, we have an entity that will be listened to and perhaps referenced, but I want to know what has been lost in the interim. Unlike the way the oral narrative works in our daily lives, oral histories are things that can take on a life independent of personal mediators (the original tellers, their interviewers, and other people who might have been present). The recording is not the original telling; it is an entity derived from the telling. The separation between narrative formation and delivery from the product on tape—or in an audio file—makes it imperative that we document as much of the original account as possible so that future users can get close to the intended sentiment. The bottom line for me is that the words alone are not enough to retell the story, and if our goal is to understand the original telling, we need to go beyond the recording and document as fully as possible the initial exchange. At the very least, this means chronicling the circumstances of the recording, the intent and interest of both recorder and teller, previous recountings of the information on the part of the speaker, and some historical and cultural context of the subjects discussed. These constitute the data that we bring, in varying degrees, to any in-person narrative exchange. We listen and respond based on our understanding of these factors as well as the actual words that are spoken. That is what we need to preserve. We repeatedly come back to the human interchange in the moment—the history and relationship between participants, the events of the day that might influence story, and the overall complexity of any communication.

I like to think that oral narratives and their power to convey information from the past are what separate us from other animals—the ability to teach through a storehouse of stories, told to one another, often from generation to generation. This is the basis of our knowledge, and was our only archive for much of our history as a species. We take it for granted in much of our lives, but we are very dependent on this form of communication for survival, and central to the process is the actual experience of hearing and telling stories. For some societies, orally communicated knowledge is the primary and most trusted source of knowledge, even when groups face contact with literate cultures or obtain full literacy themselves.[3]

Appreciation for the nuances of oral narrative (the factors surrounding the words) can remind us how oral tradition and personal narrative function, and what we may be missing when we focus only on the words preserved in a recording or on a page, as opposed to concentrating on the meaning created when we experience the oral narrative being told, considering why it was told, and to whom. For all of us (those oriented toward an oral tradition and those of us raised in print-centric societies), it is easy to lose our bearings in a world where literacy and recording devices enable us to replicate the spoken word in print and in sound files. When we rely exclusively on audio recordings and text, we can find ourselves distanced from the communication process, as well as from the relationships between those who were present and the setting of where the story was told. Despite the obvious advantages of literacy and recorders, they can blind us to intended meaning—the type of thing that we learn from long-term experience in a place, with a speaker, and with the issue discussed. It isn't a matter of what the recorder captures; that isn't the problem. The problem is what it doesn't capture but what we need to know in order to adequately understand what was said. As we look to the potential of technology to preserve and present meaning,

I think we have to continually ask ourselves what we are not capturing in the way of meaning.

Years ago, anthropologist Richard Nelson drew an analogy between oral history and photography.[4] A camera captures part of a scene. In the hands of an expert, it can capture a lot; but it never captures all of what is going on. In a similar way, oral history recordings capture the words and the intonation, the emphasis, the silences, and the tone. In the case of interviews, they can also capture the development of a line of questioning and response, maybe even a sense of the rapport between participants, but there is important information that may go undocumented, such as what went on before and after the recording, the relationships between the people present, and their responses. At best, we have a snippet.

To illustrate this point I like to use an example from my experiences at the University of the North in South Africa. The year was 1997 and the occasion was commencement. The speaker was Nelson Mandela, University Chancellor, and former President of South Africa. President Mandela chose to talk about his years working for the resistance fighting apartheid, but within his message was a plea for the students to follow reason and not blind allegiance. Behind the surface narrative, deep in the personal story, was a message to the students in the audience who had been through a tumultuous year in which they had not often followed reason. Slight differences with the administration led to protests and demonstrations, some of which led to shutdowns of the campus. It felt like the students were still fighting the revolution. Mandela never pointed the finger directly at them but rather used his own life choices to tell the students to use reason and make wise choices. I do not know whether other chroniclers will record this event the way I have. Being there, knowing the background of the university, gave me a window into part of the meaning of what he was saying. Of course, there was a lot that I didn't comprehend—symbolism surrounding farm animals and the role of political parties, for example. I realize the recording I have of the speech is but a starting point to understanding what was said, but it was a powerful testimony to the importance of background and experience in understanding a story.[5]

In a way, captured audio is like a child separated from the rest of the family. It needs support in the form of explanation, background, and context. To accomplish this, recordings need to be placed with other tellings—those already gathered and those personally experienced—in order to reveal a fuller sense of what they mean. Stories need to find their place, and once that happens, they await retelling.[6] Of course, each person comes to the recording with their own questions, experiences, and background. They will reintroduce the story when memory and occasion call for a retelling, and the very occasion they choose will reshape the story by adding another layer of meaning implicit to the new setting. The tape may sit on the shelf, or the sound file may rest on a hard drive, but the stories they hold don't stand still. They are shrouded in the "fluidity" of our lives.[7] One of the most promising aspects of the digital age is the ability to search for and access multiple accountings and track how stories are used over time.

Sherna Gluck provided an example of fluidity from her interviews with a woman organizer with the Amalgamated Clothing Workers of America. In her first interview, the woman told Sherna that she supported the men's "squelching" of the women's request for greater representation. When Sherna returned years later and asked about the woman's role in the meeting, she indicated that she had supported the women's effort in the deliberations. The reason for a discrepancy was that at the time of their initial interview, the woman thought Sherna was "sent by the union."[8] Of course, at one level we want to know what exactly was said at the meeting, but at another level we also want to know why the story is told the way it is at each telling. The total corpus of tellings and the backgrounds of each are, I believe, an integral part of oral history work. In this case, the "back story" adds important information for an understanding of both interviews. The anthropologist Elizabeth Tonkin put it this way: "I argue that one cannot detach the oral representation of pastness from the relationship of teller and audience in which it was occasioned."[9]

Early on, Alan Dundes captured the essence of what I recognize as oral history work when he spoke of text, texture, and context.[10] Text represents the body of the story, texture is the way the story is told, and context is the circumstance under which the account is given and the background needed to understand it. These three elements mark the essence of narrative analysis and performance theory.[11] Not surprisingly, this perspective on narrative comes from folklore theory and practice where the event, call it a *performance*, is the point of initial analysis and future reference. In the past 50 years, folklorists have had a large impact on the field of oral history, and I would argue they have expanded our appreciation for what is being said, and have helped us to focus on what we should be striving to preserve. This is true particularly in areas where there are culturally recognized venues for verbal expression that we may not know how to navigate due to differences in experiences, expectations, ways of doing things, and language barriers. We can easily miss the intended meaning unless we immerse ourselves in the lives of the speakers and have supporting information to nurture our understanding.

The work of folklorists in the oral history field compliments an older and concentrated focus by historically trained scholars whose emphasis is less on exploring *expression as knowledge*, than on *narrative as information* to be weighed against other sources and sifted for reliability and validity. Comparing the two approaches, the folklore/anthropological approach seeks to gather numerous versions exploring the diversity of meanings, and the historical approach seeks to gather multiple versions to find provable content and descriptive narrative that can be used to illustrate and highlight themes. The best research in oral history is done by scholars whose scope and approach encompasses both *narrative as performance*, with implicit and explicit meaning, and *narrative as data*, to be evaluated to determine empirical reality.[12]

So far, the discussion has centered on participants in the creation of the record—the narrators who tell their stories and the interviewers/recorders who are the original recipients of the accounts. Of course, researchers not only document stories, but they also use collections that are maintained in libraries, museums, and historical societies. In both settings (as co-creators and users), they see their job as interpreters of the record. Our job as oral historians does not end with our personal level of understanding, however. We are charged with finding ways to communicate that understanding to future audiences. This brings us to the two other members of the preservation and presentation team—collection managers/archivists and Information Technology (IT) specialists.

Preserving and Accessing the Record: A Team Approach

In some cases, researchers and recorders of oral history are also the collection managers, and this places them in a unique and special position to help future users understand the background and circumstances of recordings. The collection managers, by providing access, create avenues for continual exploration as new recordings are added to collections that will expand and shed further light on various topics. They provide background on each recording so that it can be placed in a historical and cultural perspective, although they usually shy away from what might seem like interpretation out of concern for influencing future researchers' conclusions. Without dipping too deeply into interpretation, there is a great deal that they can provide. Archivist and oral historian James Fogerty provides useful direction when he recommends documentation of a project's inception, goals, and design:

> The ways in which project goals are defined, topics for discussion chosen, and narrators selected should be reflected in documents created as the project progresses. It is also important to document who participated in the discussions, since they will have played major roles in shaping the project and its context.[13]

While Fogerty may not have been thinking specifically about digital access, the considerations he raises are the basis for a project's context statement, the introduction to the work, and how it was formed. Fogerty's comments are an invitation for researchers and archivists to work together to provide the necessary background and context. Consider how powerful it would be to have an interview with the researcher, a written statement about the project, photos, and video of the interview setting, and a list of the subjects discussed. These could go a ways to establishing the context as well as an understanding of why the project was initiated or why a particular interview was made. In the digital age, we can show the relationship of any number of different types of information, and we are not limited to segregation of different formats. They can be combined to build context. All of this is possible as long as the objects (photos, documents, and film) are contextually accurate to the oral narrative and as long as they are presented in a supportive role rather than a primary one. This was something we tried to do with Project Jukebox.

Together, collection managers and researchers can create a more accurate documentation and portrayal of the narrative. This is critical when it comes to comparing accounts over time and chronicling the ongoing documentation of meaning as the record is referenced and as it evolves in oral tradition and personal narrative. With the help of researchers and the public, collection managers can enrich oral history offerings with contextually accurate supporting material, ideally, from the narrator's own collection. Links to other renderings of the story or event broaden our appreciation for how and why the story is told.[14] This should be an area of *shared responsibility*, with researchers and the public providing the contextual information for each recording and the collection manager insuring that it is considered in the acceptance, arrangement, and future access to the recording.[15]

When we talk about integrating oral history holdings into the library, the most important thing is to ensure that the integrity of the oral sources is maintained so the listener can understand and evaluate the recording on its own terms. One could argue that the same conditions should apply for other formats, like photos or postcards, but what makes the care of the oral history so important is the "shared authority" held by the narrator and the interviewer, established at the time of the interview.[16] As co-creators of the record, each shares responsibility and an interest in how the recording is understood and managed. This often takes the form of a statement by the interviewer to the narrator along the lines of, "Every effort will be made to preserve the integrity of what you shared in your recording with me." The narrator's acceptance and faith in the interviewer's ability to maintain the trust is the basis of their sharing, and this is ultimately extended to the institution that houses, maintains, and makes the recording available.

For instance, when we were preparing to make the leap to web-based delivery of recordings on the UAF Jukeboxes, an extensive review was undertaken with communities and individuals. Before this, the Project Jukebox programs had been available only at standalone computer stations, and switching presentation to the web meant a major change in how the information was delivered, so we wanted everyone to be comfortable with it. Often, next of kin were contacted, first to ensure that as many people as possible were informed of the plans, and second, to be sensitive to individuals who might not want their recordings on the Web—and there were a few who chose not to have their interviews made available online. UAF's Karen Brewster researched how ethical issues were handled by other institutions, and her resulting report outlined considerations that should be followed. (That document was placed on our website as a reference.)[17] The most important thing about the process was the appreciation by community members for the care and nurturing of the recordings; that their words hadn't been taken for granted by an institution. Experiences such as this remind us that the sharing of narrative implies a trust, and that extends to the institutions that care for the recordings and how they make the recordings available. This is an ongoing responsibility.

Keeping these responsibilities at the forefront, it is time to recognize the third key player of the oral history team, the IT specialist. This is the person who designs a functional website that meets the program's need for maintenance and migratability, and serves the rest of the team's need to preserve meaning from each recording session. Through the years, in the UAF Jukebox Project, there has been a lot of discussion about what sort of people we should hire to work on the Jukebox programs. I favored anthropologists and historians because of their content knowledge, interviewing skills, and sensitivity to how meaning could be preserved, but others thought we should hire computer savvy folks so that we weren't always tearing our hair out while either trying to develop new technology or trying to get the programs to work when we didn't have the skills or background to do so. Mary Larson, who was with Project Jukebox from 1992 to 1998 (and was an anthropologist, not an IT specialist), will recall the Saturday night call she received from me in Fort Yukon, a small remote village where I was demonstrating a Jukebox program with interviews from the local community. Falling prey to the inevitable computer glitches, I begged for help. We can't remember if Mary was able to talk me through the problem or not, but the memory of the frustration is still fresh in my mind.

However, IT folks do more than rescue us when technology has us pinned down. In the age of remote delivery, they design functional ways to present oral history from a distance. While we now have an IT specialist on the Jukebox staff, our program history underscores a continued concern that technology not take over the focus. We try to emphasize how we cannot lose track of the meaning we are trying to preserve, even though we are working with machines that demand structure, order, and consistency in display and navigation.

For instance, we recognize that our oral history site has to be searchable in multiple ways, and, if the site is going to be anything more than an exhibit, it needs to be integrated with the other primary and secondary sources of the institution. The IT staff needs to insure that the oral history collection is hardwired into the institution's other holdings so that visitors to the UAF Library's website can find an oral history as they search for topics. As the site grows with more interviews, there have to be manageable ways of storing and migrating information.

At UAF, we are now in the midst of a major restructuring of how we technically deliver our programs. Historically, each of our programs was slightly different, to accommodate a look and feel for the subject. The focus was on creating an

Figure 1.1 Screenshot showing the use of physical wallpaper from the Russian Bishop's House as virtual wallpaper for the related website.

aesthetic relevant to a specific project. We used different color schemes, specially tailored backgrounds and symbols, and were especially sensitive to cultural and historical relevancy. For instance, in the Russian Bishop's House digital walking tour portion of the Sitka National Historical Park Project Jukebox, period wallpaper was used as the background on the web pages to create a sense of the actual rooms of the building (Figure 1.1). Other examples include navigation buttons shaped like fish, an otter, or a Native kayaker, and backgrounds based on Native design motifs. Our intention was to personalize and honor each project's unique features and source materials. With the number of programs expanding, this posed challenges for the library's Information Technology Department. The library IT staff insisted we develop a single and consistent Jukebox format that was easier for them to maintain and support. They needed a more efficient and economical way to manage the large and ever growing volume of digital records. Jukebox is in transition, and there is reason to worry about what we have lost by standardizing, but our reality now is to find creative ways within the structure to preserve the meaning we were attempting to capture in the earlier, more tailored programs. With technologically driven changes, we try to look back and find new ways to retain the best of what we had in a new format.

The World of Digital Possibilities

Most oral history collections are not online at this point in time, although it is probably the case that many programs are now moving toward converting their analog holdings into digital formats for ease of long-term preservation and access. (One problem is that analog cassette recorders and even older reel-to-reel recorders are about as scarce as typewriters.) We're all going digital, but how do we select which analog recordings to make digital, and then how should we arrange, contextualize, and make them accessible? There is a tendency to want to preserve the oldest and most endangered. Many of these recordings focus not so much on a topic but on an individual and the multiple topics of a lifetime of experiences. How do we weigh a collection of disparate interviews covering multiple topics against a group of interviews that focus on a particular theme that merits development, such as the stories of Holocaust survivors? There are good reasons for building online collections that develop specific topics. For one thing, most theme interviews also include other aspects of individuals' lives in the course of the interviews. A series of oral histories on a theme can also provide historical and cultural description of a topic from the points of view of different participants or observers, enriching our understanding of how an event was experienced.[18]

The selection of recordings to digitize is important, but finding the best ways to retell a story online is where the hard work begins. Because of the structure of technology, it is critical that a platform be developed with enough flexibility to allow multiple and different types of supporting information. There is a strong pull to simply create a full catalog record and provide a URL to the audio, which does provide access and meets the *traditional* library concern for integration in the catalog and consistency with other items. Time and money are some of the stumbling blocks to fuller development, but perhaps there are two larger road blocks: an aversion to anything that might look like interpretation and the archival legacy of separating different archival formats, with tapes/audio files in one place and manuscripts and photos in other locations. As collection managers race to convert old collections into digital formats, they may convincingly argue that their first order of business is to preserve the physical recording and provide the raw digital file with a library record that links the recording to their other holdings, with all else being a secondary priority. That's a defensible position, but programs that take that approach might consider creating a flexible enough delivery system so that they can, in the future, link related information to enhance the record, adding information as time and resources permit.

Now, with digital programming, we can link and draw on multiple formats to contextualize the story, and we can do this while carefully referencing the original provenience of each item (photo, text, audio, map, film). This is what we have tried to do as we have developed Project Jukebox. We provide a context statement placing interviews in historical/cultural perspective, and photographs of the speaker and the topics discussed assist future visitors to the site in appreciating what was said (Figure 1.2). As long as these images are historically and

Figure 1.2 A page from the Dog Mushing in Alaska Project Jukebox, showing a contextual statement at the top, topical navigation at left, the audio player in the center, and the transcript to the right.

Figure 1.3 A screenshot from the Pioneer Aviators Project Jukebox, showing already extant film excerpts that were digitized and included in the project.

culturally accurate to the topic and the narrator's intent, they should help lead to greater understanding. It would be a missed opportunity if the many oral history programs that are now preparing to convert their analog tapes to digital don't develop robust delivery systems to accommodate supporting information when available, as appropriate, and as time permits.

I recognize that it is always harder to get funding to work on existing recordings than to make new ones, and we certainly have faced this dilemma in the oral history program at UAF. We have had some limited success using grant funding to support the creation of new interviews together with the digitization of older recordings that address the same topic (e.g., the Pioneer Aviators Project Jukebox) (Figure 1.3). This approach has also allowed us to help fund the digitizing and indexing of relevant photos and historic film footage from archival collections. In addition, it models the coordination of a team of researchers, collection managers, and IT support, and on some projects, it has given us a chance to go back to interviewees and their families for supporting information. There are two obvious advantages to this approach. First, the oral history remains the focus of the website, while the other formats (photos, film, text) are placed in supporting roles. Second, while the volume of digitization is low (fewer than 600 recordings in all of the Jukeboxes compared to the 10,000+ recordings in the UAF Oral History Collection overall), the approach has forced us to prioritize and to develop theme-based projects that give a variety of perspectives on issues of historical interest.

There aren't a large number of programs that have tackled their entire backlog of analog recordings and developed thematically rich online records. The Virtual Oral/Aural History Archive (VOAHA) at California State University Long Beach is an exception. Under the direction of Sherna Gluck, the site represents the research efforts of students and professionals and features collections on topics such as women's suffrage, women workers in World War II, labor history, and Indian studies.

Older programs with a large number of recordings that were early subscribers to the potential of the digital age, such as VOAHA and the UAF Jukebox offerings, face the momentous challenge of converting their programming and audio-player formats to meet new IT demands, and funding for conversion has not often been part of the long-term plan. Ironically, it doesn't always pay to be the first to develop a new approach, but then, technological opportunities keep coming up. It is hard to wait for others when you recognize potential, and in this area, there is growth even when you have to make big changes later on.

For example, a big opportunity for Jukebox came at the 2007 International Oral History Association Conference in Sydney, Australia, where we met Dr. Robert Jansen of Turtlelane Studios and saw examples of how he was representing oral history using Testimony Software. This platform allows simultaneous, synchronized access to video, transcripts, a table of contents, and photographs, all on a single screen. When you click on a topic in the table of contents, the relevant section of video plays, and the transcript scrolls along under the video player. Photos pertaining to topics—or keywords discussed—change on the screen as the person speaks. This simultaneous, synchronized access was revolutionary to us and fit our goals of always wanting to provide more context and historical connections to our interviews. The implication for oral history is that you have multiple ways of experiencing the narrative and context. Our work with Jansen changed the whole look of our Jukebox sites, because we could now represent text, images, and audio/video on the same screen, with a coordinated searching or continuous-play mechanism. The problem that arose, however, was that Testimony Software was ultimately incompatible with the IT needs of the UAF Library, which wanted to work with a more ubiquitous platform that fit into the library's larger electronic-delivery goals. The Project Jukebox staff members are now attempting to replicate the functions, look, and feel that Testimony provided, while using software that can be more easily supported by the IT portion of our team. Because Testimony gave us a new look at what we could do with Project Jukebox, we have no regrets about the work we did with that software, and we can use what we learned to further develop our projects.

Lessons Learned and Opportunities to Be Pursued

I began by pointing out that the focus of oral history should go beyond preserving words spoken to preserving meaning. Background and contextual information are important for an understanding of intended meaning, but it is clear that this material needs to be appropriate to the narrative. Often this type of supporting information can best be provided by narrators drawing from their personal collections. Digital technology has the potential to effectively present this supplemental material, but only if it is organized and presented so as not to overshadow or take attention away from the oral narrative.

One of the hallmarks of oral history is the opportunities it provides for personal perspectives on issues that we may know about in a general way, but have not yet viewed through the eyes and voices of the project participants. In the digital age, where funding for collection development is hard to come by, I believe we can expect to see more theme-based oral history projects, where the project proposers can point to areas of the historical/cultural record that are under-represented in archival collections. In this environment, we see good opportunity to ensure that the speakers remain central to the delivery of information, since they can add personal perspective not usually found in other sources. There is also the opportunity with these projects to convert analog recordings from collections that relate to themes and add them in a digital format to the site.

While we have portrayed the divisions between researcher/oral historian, collection manager, and IT specialist as distinct in order to emphasize the range of necessary skills, it is also important to point out the need for a common vision of how to bring the meaning of the oral narratives to the intended audience. If the team can share a commitment to keeping the narrators' stories as the primary focus and the site visitors' main avenue into a subject, and if there is sufficient background and supporting information, the recording stands a good chance of conveying the essence of the oral narrative.

In the oral history world, digital delivery is here to stay, and the question is whether we will make the most of it. Perhaps this discussion can serve as a reference point for those embarking on the digitization of old recordings or those designing new programs. As we struggle to harness the opportunities of the digital age, we remind ourselves that our focus must remain on how to preserve meaning. Technology is a useful—but not perfect—tool for the job, and the demands and constraints it imposes must be weighed against the goal to find ever better ways to ensure a fuller understanding. As the record shows, there is good work being done in this area and much to learn from the work to date.

Notes

1. Doug Boyd, "Enhancing OHMS, Enhancing Access to Oral History," *Digital Omnium*, June 3, 2014, http://digitalomnium.com/enhancing-ohms-enhancing-access-to-oral-history/.
2. Willa Baum, "The Other Uses of Oral History," in *Sharing Alaska's Oral History*, ed. William Schneider (Fairbanks, AK: Rasmuson Library, University of Alaska Fairbanks, 1983), 38–39; Donald Ritchie, *Doing Oral History* (New York: Twayne Publishers, 1995), 1.
3. William Schneider, "Interviewing in Cross-Cultural Settings," in *The Oxford Handbook of Oral History*, ed. Donald Ritchie (Oxford: Oxford University Press, 2011), 51.
4. Richard Nelson, "The Elusiveness of Words," in *Sharing Alaska's Oral History*, ed. William Schneider (Fairbanks, AK: Rasmuson Library, University of Alaska Fairbanks, 1983), 18–19.

5. A fuller discussion of the speech is provided in William Schneider, ... *So They Understand: Cultural Issues in Oral History* (Logan: Utah State University Press, 2002), 3–7.
6. William Schneider, *Living with Stories: Telling, Re-Telling, and Remembering* (Logan: Utah State University Press, 2008), 7–10.
7. Carolyn Hamilton, "Living by Fluidity: Oral Histories, Material Custodies, and the Politics of Preservation" (paper presented at the international conference Words and Voices: Critical Practices of Orality in Africa and African Studies, Bellagio Study and Conference Centre, Italy, February 24–28, 1997).
8. Sherna Gluck, "The Representation of Politics and the Politics of Representation: Historicizing Palestinian Women's Narratives," in *Living with Stories: Telling, Re-Telling, and Remembering*, ed. William Schneider (Logan: Utah State University Press, 2008), 137.
9. Elizabeth Tonkin, *Narrating Our Pasts: The Social Construction of Oral History* (Cambridge: Cambridge University Press, 1995), 2.
10. Alan Dundes, "Texture, Text, and Context," *Southern Folklore Quarterly*, 28 (1964): 251–265.
11. Richard Bauman and Charles Briggs, "Poetics and Performance as Critical Perspectives on Language and Social Life," *Annual Review of Anthropology*, 19 (1990): 59–88.
12. Ronald Grele, "Review of *Oral History Theory*, by Lynn Abrams," *Oral History Review*, 38(2) (2011): 355.
13. James Fogerty, "Oral History and Archives: Documenting Context," in *Handbook of Oral History*, ed. Thomas Charlton, Lois Myers, and Rebecca Sharpless (Walnut Creek, CA: AltaMira Press, 2006), 211.
14. Paula Hamilton and Linda Shopes, *Oral History and Public Memories* (Philadelphia: Temple University Press, 2008).
15. Schneider, ... *So They Understand*, 161–168.
16. Michael Frisch, *A Shared Authority: Essays on the Craft and Meaning of Public and Oral History* (Albany: State University of New York Press, 1990).
17. Karen Brewster, "Internet Access to Oral Recordings: Finding the Issues," last modified October 25, 2000, http://library.uaf.edu/aprc/brewster1/.
18. Carrie Nobel Kline, "Giving it Back: Creating Conversations to Interpret Community Oral History," *Oral History Review*, 23(1) (1996): 19–40.

5. A [illegible] of the speech is preserved in William [illegible] [illegible] The [illegible] [illegible] [illegible] (Georgia State University [illegible] [illegible]).
6. [illegible]
[illegible]
7. [illegible] the [illegible] of [illegible] [illegible] the [illegible] [illegible] [illegible] [illegible] [illegible], in [illegible] and [illegible] [illegible] [illegible] [illegible] [illegible]
8. [illegible] "The Representation of Politics and the [illegible] [illegible] [illegible] [illegible] [illegible] [illegible] [illegible] [illegible] [illegible] [illegible] [illegible]).
9. [illegible] [illegible] (2011).
10. [illegible] "[illegible] [illegible] and [illegible]," [illegible] [illegible] [illegible]
11. [illegible]
12. [illegible]
13. [illegible]
14. [illegible]
15. [illegible]
16. [illegible]
17. [illegible]
18. [illegible]
19. [illegible] (1996): 17–40.

CHAPTER 2

Why Do We Call It Oral History? Refocusing on Orality/Aurality in the Digital Age

*Sherna Berger Gluck**

Honoring Orality before the Digital Revolution

When I set out to record women's experiences in 1972, my commitment to "recovering our past" had little to do with the oral history movement and everything to do with the women's liberation movement. Like so many other feminists around the country who were engaging in this avowedly advocacy-oriented work, when we launched the community-based Feminist Oral History Research Project (FHRP), we had just enough money to buy some audio cassettes and pay for a post office box. We certainly did not have the money or womanpower to create transcripts. Besides, at that time, we were not primarily interested in creating an archive; we were about giving voice.

The mantra of the late 1960s/early 1970s US new social historians, and of the second-generation US oral historians among them, was "giving voice to the voiceless." That had a nice ring to it, but feminists did not believe for one moment that women had been voiceless. Rather, their voices had not been heard; they had been muffled, muted, or drowned out. Our emergent generation of feminist oral historians was determined to change that and to make women's voices heard—not just their words on the printed page, but their unmediated *voices*. In other words, we wanted to preserve the complex performance of our oral history narrators with their inflections, pitch, pace, and rhythm instead of flattening these to a monotone in a transcript. We were on our own trajectory, and I doubt that any of us were aware of the tape vs. transcript debate raging in the oral history movement. Nevertheless, our instincts were not beyond the pale and were in line with the sentiments expressed by (UK) History Workshop founder Raphael Samuel, who had noted how the "spoken word can very easily be mutilated when it is taken down in writing and transferred to the printed page."[1] More than that, the nuances of the performed narrative are completely lost no matter how many *stage directions* are inserted into a transcript.

In any event, at least initially, we were more interested in collecting oral histories and in having the women's voices heard by others than in processing the recordings. In 1974, for instance, the FHRP produced a tape-slide show, *Recovering Our Past: The Struggle for Women's Suffrage*, which used excerpts from its suffragist interviews, and we later distributed an audio documentary on reproductive rights titled "*I Tried Everything, But Nothing Worked: Birth Control in the Early 20th Century*." Obviously, the technology of the 1970s made distribution of these programs very limited. Now, however, these early multimedia ventures featuring voice can reap the benefits of the digital revolution. Indeed, the suffrage tape-slide show can now be widely accessed online.[2]

Nevertheless, as we were engaged in these early efforts to deliver the voices of women, it was becoming clear that we were indeed creating valuable archives, although these often were little more than tapes stored in file drawers in our homes or in our community women's centers. Understanding the need to share this new primary source and make it usable, some of us devised summarizing techniques that facilitated listening to the analog recordings.[3] Others, inspired by anthropologist Dennis Tedlock, tried to capture the performative nature of the oral history interview through the use of free verse and creative formatting of the text.[4] Though limited, this creative effort at least enlivened the printed version of the spoken word.

Our various summarizing methods facilitated the use of our oral histories, and the FHRP interviews with labor activists, in particular, were used by quite a few scholars, even before they were deposited in an archive.[5] Nevertheless, it was both cumbersome and time consuming to locate and listen to portions of an analog recording, especially in contrast to flipping through the pages of a transcript. Furthermore, transcripts provided an easier source for publications.[6] For all these reasons—and especially because the printed word was the *raison d'être* of libraries, where most early US oral history programs were housed—the debate was settled in 1977 in favor of the printed word. Although favoring the transcript meant abandoning a commitment to orality, Louis Starr somewhat presciently noted that "future generations might be more aurally oriented."[7]

Indeed, not only is there a more aurally oriented generation, but we now have the technology to operationalize a commitment to orality and its complement: aurality. Furthermore, through the World Wide Web, we are able to deliver on the very democratic impulse of the oral history movement by making the oral histories that we collect widely available for the use of others. We have to be mindful, however, that the potential to deliver oral histories on the web, particularly in their original oral form, is not without its own problems, and even perils.

Entering the Digital Age—First Questions, First Steps

When Bret Eynon, Donald Ritchie, and I reflected on "Oral History in the New Millennium," the landscape had already changed. Unsurprisingly, without prior discussion among us, we all independently focused on the impact and potential of new media, and particularly the World Wide Web, although we didn't all concentrate on the potential for honoring orality.[8] By then, I had been addressing both my quirky fears and legitimate ethical concerns and already had a glimpse of the possibilities of the digital revolution.

A video artist helped quell my fears about the potential to lift, morph, and manipulate the voices we might put online for other, perhaps even malicious, purposes. "After all," she asked, "how is this any different from the ways that the printed word can be distorted, decontextualized, and misused?"[9] I still wonder, though, if voice isn't different; if it isn't more deeply a marker of identity than a printed rendition of our words might be. Perhaps this is a primeval fear akin to the belief of some peoples that photographing them steals their soul. I cannot vouch for the safety of the soul, but discussions among colleagues in the oral history community helped to reassure me that we could craft ethical solutions.

Mary Larson, originally from the University of Alaska Fairbanks Project Jukebox, helped pave the way for me, frankly discussing some of the solutions that they devised for their groundbreaking project.[10] While some of their solutions, like distribution through LAN (Limited Access Network) were unique to their context and their particular sensitivity to the Native Alaskans, with whom they were engaged, this practice pointed to a range of possibilities. The intense debate in 1998, first among members of the Oral History Association (OHA) sub-committee—charged with revising the OHA *Evaluation Guidelines*—and then in the association's business meeting where the committee's recommendations were introduced, convinced me that it might be possible to navigate the ethical fault lines.[11] As I will discuss later, in trying to implement these ethical guidelines for oral history on the web, we frequently were given a wonderful opportunity to connect with narrators whose oral histories had been deposited as much as 25 years earlier.

Serendipitously, at that same 1998 OHA conference in Buffalo, I saw a demonstration of the Interclipper indexing program, which enabled segmenting and summarizing of segments while simultaneously digitizing the analog tape. The following year, I brought some sample analog tapes with me to the Anchorage conference and sat in the exhibit room testing the methodology. I was as delighted as the other oral historians who gathered around the computer to watch. It seemed like the perfect one-stop solution for summarizing and digitizing the more than 1,000 hours of oral histories that were by then deposited at California State University, Long Beach (CSULB).[12]

Armed with what seemed like the perfect mechanism to realize our goal of preserving the audio and presenting and delivering oral histories in their original form, Kaye Briegel and I initiated a pilot project in the fall of 1999, with seed funding from the College of Liberal Arts and the University Library. Creating organic segments and summarizing and assigning keywords to them based on the flow and content of the narrative was certainly preferable to the arbitrary three-minute time segments we had used in the analog age. In very short order, however, it became obvious that Interclipper's ability to digitize and summarize organic segments simultaneously was a major drawback. Its design, at least at that time, meant that there was not a digital file of the entire recording. Furthermore, at that point, the resulting Interclipper database was not transferable to the web. Because we wanted to use the potential of the new technology to its fullest, both to preserve the integrity of the original recordings and to make our valuable primary audio documents available on the web, it became apparent that this proprietary software did not suit our needs.

Thus began the collaboration between the oral historians and the technical staff of the Academic Computing Services to develop more appropriate methodologies for oral history delivery on the web, and ultimately, the creation of VOAHA: the Virtual Oral/Aural History Archive.[13] First, we discussed how best to digitize the analog recordings and what formats were most appropriate for preservation.[14] Beyond that, however, we dialogued on how to develop the systems necessary to process and deliver the oral histories in a form that met scholarly requirements. Because of a shared commitment to scholarly standards and broad disciplinary backgrounds, the technical staff was as likely as the oral historians were to come up with suggestions that furthered the scholarly value of the site. For instance, the development of an automated citation for each interview segment was triggered by a discussion with the librarian on the team.

The development of the original VOAHA site was an ongoing process, and whenever the oral historians came up with new ideas—for example, introducing photographs—the technical staff worked out the best means for delivering the material in conjunction with the audio files; and whenever the technical staff came up with new ideas for presenting associated information, we oral historians assessed how it could best serve other practitioners.

The collaborative process, which initially also included the participation of the Women's Studies and Ethnic Studies librarian, was not always smooth sailing. The value of some of the oral historians' requirements was not always self-evident to the technical staff (for instance, the need to fully contextualize the materials, including providing information on the interview *process*). Cognizant of the technology and the advanced tools for web scripting, the technical staff wanted to deliver the material in the most immediate way, namely, getting to the narrative *summaries* and the tape segments as quickly as possible. After engaging in substantive discussions, we reached compromises that satisfied all of us and resulted in a design that was both academically sound and technically innovative in linking text, images, and audio.[15]

This meant that each oral history on the site was fully contextualized and was grounded within both its series and the larger historical context that was elaborated upon in the collection description. An automated built-in citation format for each segment, which also registered the date the segment was accessed online, was a great addition to the documentation provided for each interview.[16] While this kind of documentation is relatively straightforward and more *objective*, we also were committed to including the interviewers' subjective impressions of nonverbal cues, the nature of the environment in which the interview was conducted, and the factors that might have affected the relationship between the interviewer and the narrator. One stark example suffices to illustrate the critical role that subjective impressions can play in shaping a narrative.

In 1975, when I first interviewed former union organizer Mamie Santora, I was particularly interested in hearing her account of the contentious 1926 convention, where rank-and-file women were pushing for a Women's Department. Santora was the only woman on the executive board at the time, and the written record seemed to indicate that she supported the male leadership's "squelching" of the proposal. Indeed, when I first interviewed her, Santora said that she agreed with and supported the position of the male leadership. However, when I went back to see her several years later and returned to the topic, she "spilled the beans." As it turned out, she had thought earlier that I had been sent by the union. When she realized that this was not the case, she was more candid and less cautious in her criticisms of the union and revealed the dynamics that led to her own vote and the defeat of the resolution. This example illustrates not only how a changed understanding of the interviewer's positionality affected the narrative, but also how the changing of the political context and historical moment during which an interview is conducted are part and parcel of the dynamic. In fact, it is hard to extricate one from the other.[17]

To the extent possible, all these elements of the interview process were incorporated in VOAHA into a hierarchical structure that started with the collection description and ended up with the segment summary from which the audio was directly accessed. This enabled the more casual users, such as students searching for a topic, to browse starting with the list of collections and series (Figures 2.1, 2.2), ultimately working their way down to the narrator(s), and the interviews that interested them. They could then listen to the entire audio file or to specific segments. On the other hand, the more focused, deliberative user was able to do a direct search, either by selecting a term from the index of keywords (names and subjects) or by entering a wild search term in the search box. Either of these search methods would bring up a list of *every* audio segment *across the entire 1,000 hours* of oral histories that matched the search term, as a result of either an assigned descriptor or the appearance of the term in a segment synopsis (Figures 2.3, 2.4, 2.5). Ultimately, we realized that this wonderfully productive search process often meant that the user did not see all the carefully constructed contextual material. We believe that this problem can be addressed by a design

Figure 2.1 Screenshot of VOAHA, Women's History Series List.

Figure 2.2 Screenshot of VOAHA, Interviewee List of (Suffragists).

Figure 2.3 Screenshot of VOAHA, Interviewee.

Figure 2.4 Screenshot of VOAHA, Interview (Bio).

Figure 2.5 Screenshot of VOAHA, Individual Segment.

where references to all the contextual materials are displayed on the page rather than needing them to be accessed through a step-by-step sequence.

Despite a drawback like this that became obvious only after the fact, like the earlier University of Alaska Project Jukebox, in 2000, VOAHA was ahead of the curve. With no ready design or architecture on which to draw, this highly interactive *boutique* site was recognized for its innovation and was awarded an Accenture/MIT Digital Government Award in the Higher Education Innovator category in 2003. As discussed later, it would be seven years before it became necessary to grapple with the institutional and infrastructure challenges that threatened the very survival of VOAHA.

From Mechanics to Ethics

At the same time that the technical team was working out the design and mechanics of the original VOAHA site, the oral historians were grappling with how to implement the well-intentioned but somewhat vague ethical prescriptions for putting oral history on the web. Most of the oral histories deposited at CSULB followed the simple FHRP agreement that made the material available for educational and research purposes.[18] Since these agreements did not cede rights to heirs and, by and large, did not impose any limitations other than the occasional request for anonymity, the constraints we faced were ethical rather than legal.[19]

Following the prescription adopted by OHA in 1998 to make a best-faith effort to obtain agreements from narrators before mounting their interviews on the web, we first tried to ascertain what narrators were likely to still be living, based both on their approximate age and on available records.[20] Because the first batch of women's oral history narratives to be launched had been conducted when the narrators were already in their eighties and nineties during the early 1970s, they weren't likely to still be alive. As a result, with no further obligation to heirs and with the narrators' stated intentions to have their stories told, we processed these without hesitation. On the other hand, the narrators of the 1973–1974 recordings about life in the Japanese fishing village on Terminal Island were in their twenties or even teens when they were forced to evacuate in 1942, so many were likely to still be alive in the late 1990s.

Indeed, because this tight-knit Terminal Island community reestablished itself as a social community after the war, it was not too difficult to make good-faith efforts to contact many of the Nisei narrators. Taking a lesson from Project Jukebox, we offered to visit them and demonstrate how the site worked. Some were eager to see the demonstration; others were simply pleased that we had contacted them and willingly sent in photographic materials. Without exception, they were delighted that their story would be told. Given this uniform reaction, when my best efforts to locate some of the survivors failed, I felt confident that the spirit of their intent was being honored.

Even so, a more nagging and thorny issue looms large, namely what constitutes informed consent when we are talking about a quantum leap in distribution via the World Wide Web? Could the elderly Nisei who generally did not have familiarity with the Internet conceive of what it would mean for millions and millions of people around the world to be able to access their oral histories? For these narrators, the significance they attributed to recapturing the life of a community that was destroyed by wartime hysteria and xenophobia outweighed any reservations that we might have had.[21]

Flush with this success, I nevertheless remained concerned that the much younger veterans of the late 1960s and early1970s Los Angeles women's liberation movements might be reluctant to have their oral histories go so very public. At the time that many of us were interviewed in the 1980s, we still subscribed to the belief that the "personal is political," and usually did not hesitate to bare our private lives. To my great relief, people were enthusiastic rather than reluctant. In fact, five of the narrators eagerly sent revisions to the biographical sketches we drafted, and one even worked on revising the series narrative for the Asian American women's movement.

Although we were allies and comrades, I believe that their level of engagement goes beyond our personal relationship and is somewhat akin to the Terminal Island Nisei's appreciation of having their story told. Until the very recent challenges to the historiography of Second Wave feminism, the legacy of southern California activists of all colors and backgrounds was ignored. Most histories had focused mainly on the East, Midwest, and northern California, with the result being the virtual invisibility in the record of some of the first Chicana and Asian feminist organizing.[22]

Beyond Permission—the Implications of the Quantum Leap in Distribution

There is no question but that placing an oral history on the web is different from earlier forms of distribution, even published books. Indeed, it is a quantum leap, and the resulting public scrutiny raises issues that we must still consider, regardless of the narrator's agreement and/or intent. In fact, consequences of both the personal and political implications of web distribution and the potential for unintended uses by others might even make us reconsider how we conduct our interviews.[23]

The troublesome implications of wide distribution, especially coupled with powerful search capabilities, also have to be understood within the context of changes in both the life circumstances of narrators and sociopolitical environments. These issues emerge not only as relics of the kind of revelations recorded in a different time and culture, but are still pertinent today. Kelly Anderson, Smith College oral historian, notes that she is growing reluctant to put oral histories online from the Sophia Smith Collection's "Voices of Feminism" and "Documenting Lesbian Lives." Noting that narrators have sometimes revoked their permission for online access after Googling themselves, she observes, "The material no longer feels like an archival document with any kind of gatekeeper but rather a trove of personal information available to the masses."[24]

In addition to the issue of potential personal embarrassment of narrators, we must also ask what the political implications might be of making some material so widely available with the click of a mouse or the tap on a screen. In today's increasingly repressive and intrusive political climate in the United States, concern about political implications is more than an idle hypothetical issue.[25] Does this then mean that we should go beyond acquiescence and discuss the possible ramifications of their oral history being accessible on the web with surviving narrators? And if we can't locate them, must we take responsibility for making decisions that take full account of these implications? Furthermore, today, how might advance expectation of web access influence the interview process? I turn to the golden rule of interviewing that I always advised students: "Never ask a question that you are not yourself willing to answer." Today I would add "that you don't want the world to hear via the web." Therefore, although I had no hesitation in answering questions some 16 years ago when I was interviewed about the women's liberation movement in Los Angeles, today I might very well be less forthcoming about experiences that might be personally embarrassing, or especially activities that might place me at risk politically. In other words, in the future, I might well monitor narrators—and myself—more than I have in the past.

Of course, narrators were cautious even before the possibility of such wide, unmediated access, and they often refused to discuss some matters on the record. If I felt it was of sufficient historical importance, I would try to negotiate with them to find a way to talk about the material. The prime example of the success of this kind of process is the interview with suffragist Laura Ellsworth Seiler. In talking about the Women's Political Union, Seiler was initially reluctant to discuss the authoritarian leadership style of Harriet Stanton Blatch. However, after I noted the historical importance of her observations, she agreed to allude to Blatch's leadership style, but without many details. That was sufficient, however, both to provide some critical information to Ellen Dubois, who was researching the organization, and for one of her students to accuse me of excising material in a second edition of *From Parlor to Prison*. In other words, there was enough information implied between the lines that the student was convinced that there had been more detail.[26]

Self-censoring by narrators is one thing, but now, with the possibility of unmediated web access, we must think about our responsibilities as interviewers. Should we be more cautious in what we ask, or even caution narrators, reminding them about the implications of wide, public distribution? I am inclined to believe that we should do just that, even though this might well stifle the spontaneity of an interview. If nothing else, we certainly should negotiate recording of materials in a way that takes into account the implications of easy access and broad distribution over which we have no control. Even if the user accepts whatever agreement might be required to enter a website, this is not an enforceable agreement, and even if the audio files are not downloadable, there are simple ways to work around this. Even with the early meager efforts at control of access that both VOAHA and Project Jukebox instituted, their newly migrated sites at their respective libraries have, at best, a simple copyright notice at the bottom of the page. In other words, there is no longer a use agreement or any other control over access.[27] This lack of control over access, coupled with the potential for personal embarrassment and/or political repercussions, leads me to an unhappy conclusion; namely, if we want to behave responsibly and ethically in placing oral histories on the web, we have to be prepared for the tradeoff that this might entail.

What a Difference a Decade Makes: Institutional Changes and Infrastructure Obsolescence

In the years since VOAHA—and Jukebox before it—developed its own methodology, technology has grown by leaps and bounds. As a result, these projects are now behind the curve. Additionally, both institutional and infrastructural changes have necessitated migration to new architecture. Below, I will discuss this process and also suggest how important it is for oral historians to develop recommendations for best practices of oral history websites.

While every oral history program and web project is different, and while VOAHA's experience in the last decade might be unique in some ways, there are important lessons to be learned with regard to both the issue of institutional support for web projects and the effect of rapidly changing technology. Not unique to it, however, is the impact of budget slashing on universities and colleges, public programs, and governmental funding from agencies like NEH (the National Endowment for the Humanities). Our experience with VOAHA points to serious questions about both where and how projects are institutionally located and even their ability to be assured ongoing support, especially in educational institutions whose primary mandate is teaching.

From its inception as the Oral History Resource Center in 1978, the program at CSULB was not designed as a collections program. Rather, it encouraged, trained, and assisted faculty and students to develop projects. These were usually deposited in the Special Collections and University Archives, joining the FHRP collection and the Terminal Island and Mexican American oral histories. Through special funding and individual grants, several projects were coordinated out of the center, which was then a freestanding program in the College of Liberal Arts. This included the Rockefeller—and NEH-funded Rosie the Riveter Revisited and the Long Beach Petroleum projects. While still retaining its essential role as an instructional program, during the budget crunch of the early 1990s, the OHRC was scaled back and incorporated into the History Department, with only part-time funding for the director and a student assistant.[28] As a result, when VOAHA was conceived, there was no clear, funded, institutional base.

VOAHA, therefore, was structured as a collaborative venture of three divisions of the university: the Library, the College of Liberal Arts (CLA), and the Division of Information Technology's Academic Computing Service. A pilot project, initially funded by the Library and CLA, helped to garner a grant from a local foundation to launch 300 locally focused oral histories on the web. Eventually, on the third try, in 2002, an NEH Preservation and Access grant funded the addition of another 650 hours. It is interesting to note that the ultimate funding reflected the changing thinking, both about the necessity of transcripts and the permissions required for placing oral history on the web.[29]

The loose, collaborative structure of the VOAHA project had a tremendous advantage at the time, enabling the faculty and staff from these three units to work creatively outside the usual bureaucratic, structural confines. Ultimately, however, the changes in the IT division, including the virtual end of the support for special faculty projects, meant that VOAHA became somewhat of a bouncing ball following a catastrophic system failure in 2010. Even before then, as most of the IT services were transferred to the Library, a proposal was made to migrate

VOAHA to what essentially was a static archival architecture. Because this might have meant the end of VOAHA's interactivity—one of the main features cited in the 2003 MIT-Accenture award[30]—including the ability to access it through search engines, this plan was rejected. Everything was placed on hold, especially with the long illness and death of the creative VOAHA designer and the retirement of the other key staff member.[31]

The catastrophic crash in 2010 of VOAHA and another NEH-funded website that was designed by the same technical team revealed a host of fault lines. These included a question of who was responsible for ongoing support; the challenges of a design that depend on programming that required specialized skill (WebObjects); and the obsolescence of some of the hardware and software that were integral to the website design. The question of responsibility for maintaining the website after the grant expired became an issue in which two divisions of NEH became involved, and points to the absolute necessity of establishing policies and guarantees for ongoing support of a website project. In fact, our experience ultimately led NEH to realize that it needed to spell out more explicitly its expectations for ongoing institutional support of projects that had been funded by the agency.

Even with the various institutional realignments and the almost exclusive focus on direct teaching-related projects, the university ultimately recognized VOAHA's national standing, as well as the role it played in our own curriculum, and brought in an outside consultant familiar with WebObjects. Although the site was eventually restored, the process highlighted some of the problems inherent in its design as well as the obsolescence of both the software and hardware on which the site depended. This included not only the outdated version of WebObjects and its server—for which, parts were no longer available—but also the approaching end-of-life of the RealNetworks server that delivered the audio (.rm) files. Because there was no guarantee that the site could or would be restored again in its original form if there was another crash, it was too risky to make any changes to it or to add new materials. So, although the VOAHA team initially resisted migrating to a completely new architecture and tried to seek alternative fixes, it became increasingly apparent that there was no other option.

VOAHA II

Both the College of Liberal Arts and the Library remained committed to preserving VOAHA, but following the restoration of the site after the fatal crash in 2010, its future remained in limbo. Finally, in 2011, with realignments in the university's IT division, it was incorporated into the Library, its future assured within the development of the Symposia technical infrastructure the university adopted for its Digital Institutional Repository.

Planning for migration to the Content Pro IRX library management software meant first archiving the entire VOAHA site (which was backed up on an offsite server) and converting the audio files to the current .mp3 format as well as preserving them in the universal AAC (Advanced Audio Coding) format.[32] With archiving and audio conversion accomplished, a crosswalk map was then created between the existing Oracle database and the requirements for the new product, a process that entailed mapping the Oracle database contents into Dublin Core metadata descriptive fields. Once the library technology staff understood how these two data sources were related, testing began with the data being manually populated into Symposia.[33] The three-month process began with writing queries against Oracle in order to create the required data and then testing, reviewing, and comparing the results to ensure that the data was complete. A program then was written to output data in XML/RDF so that data files that contained the structure, organization, and files based on Dublin Core metadata could be batch loaded into the Symposia product.

Unlike the earlier close collaborative engagement between the oral historians and the technology staff in the initial development of VOAHA, the staff involved in the migration process worked independently, in direct consultation only with the vendor's archivist. As a result, when the VOAHA oral history team saw the first iteration of the new Symposia site, it became obvious that the migration team had no understanding of the relationship between the various elements of the oral history process.[34] For example the description of the interview process for the first interview was the only one that was displayed, regardless of how many different interview sessions there were; and the list of topics for each part of an interview session's recording (originally designated as sides of the tape) were all lumped together. Following this critical review, the technology team worked with the vendor's archivist to correct these problems, rendering VOAHA II ready to be incorporated into the new University Digital Repository.

With the migration to this new platform, VOAHA has become one collection, or *community*, within the university's digital repository, and is linked seamlessly with the library catalog. In other words, VOAHA's integration into the institutional repository, and its incorporation into the library community, assures its future. On the other hand, the very fact that a library systems model—with its historic grounding in text and linearity—shaped the new VOAHA design creates a serious contradiction. In effect, the *primacy* of orality/aurality has been supplanted and the complex interactivity that made both VOAHA and Project Jukebox so innovative has been largely lost.

For one thing, while the time segments of an interview are noted in the text summaries, there is no direct link between an interview segment and the audio it is describing.[35] Rather, the user must go to a navigation bar and manually locate the audio. Even more problematic is the loss of easy comparability between different oral histories. In the original VOAHA design, each time segment was assigned descriptors or key words. As a result, a search for a topic would bring up each and every *specific* location—across the entire 1,000-hour archive—where that topic was discussed. Now, a list of narrators names might be brought up on a search if the specific term is in any of the texts, but the user will have to search within each separate narrator's interview(s) to find the relevant section and, again, must then manually navigate to find the precise section in the audio file.

On the flip side, but of considerably less importance, one of the problems noted previously about the way VOAHA was designed has been somewhat resolved with the Symposia design. Instead of going from web page to web page as the user drilled down through the elements of the oral history (the narrator's biographical sketch, the description of the interview process, and the list of topics covered in the interview)—a design that meant the user did not see all the contextualizing materials—much of the material is now presented sequentially on the same page. Specifically, the page that displays the contents of an interview includes not only the full contents of an audio file, but also the narrator's biographical note, the description of the interview process, and the list of topics covered. Most critically, because the complete contents of the audio file are displayed, the user sees the context in which a specific segment is embedded.

Migrating from an existing site, particularly one that was necessarily uniquely designed, obviously has its limitations. Although not yet achievable, we are assured that additions to VOAHA can be made. Because the tools being used by the university are constantly evolving and being adapted to changing technological environments, we hope that further evolution of the software will also be better able to accommodate some of VOAHA's creative design and restore the primacy of orality/aurality.

VOAHA and Project Jukebox might be casualties of their early innovation, which necessitated designing sites out of whole cloth with whatever tools were available. Today, however, with the proliferation of oral history websites, the need for some universal applications has become apparent; and fortunately, this is now being addressed. Similarly, more attention is being paid now to best practices for oral history on the web. While these might vary with institutional infrastructures and budgets, hopefully, our experience with our early projects can at least point to some of the ethical, methodological, and practical issues to be considered as we enter the next stage of our digital revolution.

Let us hope that the next stage opens up more possibilities for the complete interactivity and contextualization that so-called boutique sites like VOAHA and Project Jukebox were able to incorporate. In other words, to quote Will Schneider, "let us hope that we aren't strait-jacketing recordings into a system designed for print and with a legacy so ingrained in literacy that it only awkwardly accommodates multi-media."[36]

Notes

*The ideas presented here are largely a result of the close collaboration with my Virtual Oral/Aural History Archive (VOAHA) partner, Kaye Briegel, who was involved from the start in its development and evolution. Ali Igmen, our successor following our respective retirements, joined in our discussions later, especially after the fatal crash of VOAHA in 2010, and in our conversations regarding migration to a new architecture. Their comments on this paper are greatly appreciated. David Bradley and Nancy Rayner were the key technical staff who crafted the process and design of the original VOAHA, with the participation of Walter Gajewski who served as an advisor on audio issues. The Academic Computer Services unit in which these staff members worked was supplanted by Academic Technology Services in the Library. Technological Strategist Jill Horn was responsible for the migration of VOAHA.

I am especially grateful to my University of Alaska Fairbanks colleagues, Karen Brewster and Will Schneider, whose insightful feedback enabled me to make meaningful comparisons between our two projects and especially helped me to dig deeper and think bigger.

1. Raphael Samuel, "Perils of the Transcript," *Oral History*, 1(2) (1972): 19. Unfortunately, in the early days, some programs not only produced heavily edited transcripts, they even recorded over the tapes.
2. Thanks to William Schneider's long-standing appreciation of this early tape-slide show, it was digitized by the Alaska and Polar Regions Department of the University of Alaska Fairbanks in 2009 and is available on the Project Jukebox website at http://jukebox.uaf.edu/suffragists/narrated_slideshow/HTML/testimonybrowseraudioonly.html.
3. For instance, quite independently, both Dale Treleven and I developed a summary format for the Wisconsin Historical Society and the Feminist History Research Project, respectively. His was to create a summary in five-minute time segments; mine used three minutes.
4. Dennis Tedlock, "Oral History as Poetry," in *Envelopes of Sounds: The Art of Oral History*, ed. Ronald Grele (New York: Praeger, 1991), 106–125; Karen S. Harper Wilson, "In Hmong Mountain: See Lee's Oral Narrative as History and Poetry" (master's thesis, California State University, Long Beach [CSULB], 1995).
5. Eventually, the FHRP collection was deposited at CSULB and became the basis for several of the Virtual Oral/Aural History Archive (VOAHA) series in women's history.
6. For instance, the FHRP suffrage interviews were incorporated into the Regional Oral History Office (ROHO) Suffrage Project in order to include interviews with "non-elites." These transcripts enabled me to edit and publish the oral histories in *From Parlor to Prison: Five American Suffragists Talk About Their Lives* (New York: Vintage, 1976; reprinted, New York: Monthly Review Press, 1985). Similarly, a decade later but still before the digital revolution, transcripts were produced for the Rosie the Riveter Revisited project in order to facilitate wider distribution to various institutions. In, turn, these became the basis for my book *Rosie the Riveter Revisited: Women, the War and Social Change* (Boston: G.K. Hall, 1987, reprinted by Plume: New York, 1988). Today, the voices of these women can be accessed on the VOAHA at http://www.csulb.edu/voaha.

7. Louis Starr, "Oral History," in *Encyclopedia of Library and Information Service*, Vol. 20 (1977), reprinted in *Oral History: An Interdisciplinary Anthology* (Walnut Creek, CA: Altamira Press, 1996), 42–43.
8. Sherna Berger Gluck, Donald A. Ritchie, and Bret Eynon, "Reflections on Oral History in the New Millennium: Roundtable Comments," *Oral History Review*, 26(2) (1999): 1–27.
9. Conversation with video artist Abbe Don, who had created the "Bubbe's (Yiddish for grandmother) Back Porch" website, 1996.
10. This was over breakfast at the Oral History Association (OHA) conference, the date of which neither of us recalls, but most likely sometime between 1996 and 1998.
11. The outcome of these debates was reflected in the guidelines that were adopted in 1998 and published two years later. *Oral History Evaluation Guidelines*, Oral History Association Pamphlet Number 3, 2000, available online at http://www.oralhistory.org/about/principles-and-practices/oral-history-evaluation-guidelines-revised-in-2000/.
12. After I joined CSULB and established what was then called the Oral History Resource Center (OHRC) in 1978, I surveyed faculty to ascertain what oral histories might have been done by them and/or their students. The initial oral histories from the Asian American and Mexican American faculty—added to the FHRP oral histories and the Long Beach University and community oral histories that were being conducted by my colleague and eventual VOAHA co-founder, Kaye Briegel—became the basis for a growing interview repository in the library's Special Collections and University Archives. By 1998, especially with the ongoing collection of community and women's oral histories, including the Rosie the Riveter Revisited project, the repository had expanded. Unfortunately, the extensive interviews on musical developments in southern California that had been conducted earlier by Clare Rayner and Bill Weber were in such raw form (original reel-reel tapes with incomplete documentation) that they remain unprocessed on the shelves of Special Collections and University Archives.
13. Kaye Briegel and I were the oral historians. The Academic Computing Staff included David Bradley, who designed and created the scripts; Walter Gajewski, who advised on audio preservation; and Nancy Rayner, who designed the databases for creating and summarizing interview segments and other contextualizing materials.
14. Because the campus used a RealNetworks server at the time, the files prepared for audio delivery were saved in the .rm (Real Media) format. These were compressed from the .wav (Waveform Audio File Format) files. Additionally, audio CDs were made for each interview.
15. The technological decisions necessarily were based on both the servers and software licenses used by the university, including Oracle, and on the technological expertise of the staff. As a result of the latter, WebObjects was used to write the script, and SMIL (Synchronized Multimedia Integration Language) files were created to link the various elements.
16. A sample citation might read: Santora, Mamie. Interviewed by Sherna Berger Gluck. May 20, 1975. Labor History: Garment Workers. *The Virtual Oral/Aural History Archive, California State University, Long Beach.* Interview 1a Segment 1 (0:00–1:01) Segkey: gws5000. March 14, 2012, http://www.csulb.edu/voaha.

17. I discuss this further in "A Conversation with Sherna Berger Gluck and Ted Swedenburg," which is appended to my article, "The Representation of Politics and Politics of Representation: Historicizing Palestinian Women's Narratives," in *Living with Stories: Telling, Re-Telling and Remembering*, ed. William Schneider (Logan: Utah State University Press, 2008), 136–137.
18. The new agreement forms provide an opportunity for the narrator to opt out of electronic distribution.
19. The issue of intellectual property rights was raised by a member of the National Endowment for the Humanities (NEH) Council during deliberations on our second grant proposal. A subsequent 26-page appendix convinced the historians and archivists that we were within our legal rights to mount the oral histories on the web.
20. In the United States, death can usually be confirmed through Social Security records for those who were members of the labor force. For women who had not engaged in wage labor, the death records of the state can be consulted, although these are not always accurate or current.
21. Members of the community not only cooperated on a recent documentary, but they raised funds to erect a massive monument on Terminal Island. The commemoration of the monument in July 2002 attracted 400 members of four generations who came from all over the United States.
22. Even though this history has become more inclusive in the past couple of years with groundbreaking work on Chicana feminisms and black women's organizing, West-coast Asian American women's organizing during the 1960s and 1970s remains largely ignored.
23. For a detailed discussion, see Gluck, "Reflecting on the Quantum Leap: Promises and Perils of Oral History on the Web," *Oral History Review*, forthcoming.
24. Email to H-ORALHIST@h-net.msu.edu (listserv), July 6, 2012, Subject: Online access to LGBT oral history projects.
25. In the past several years, we have witnessed a growing number of FBI raids of activist homes, confiscation of computers, grand jury subpoenas, and even 38 and 40-year-old charges being renewed (e.g., the SF 8 [Black Panthers] and Los Angeles Chicano activist Carlos Montes).
26. Gluck, *From Parlor to Prison*, 1985. To hear the segment, go to http://www.csulb.edu/voaha. Enter "Laura Ellsworth Seiler" and go to Interview 2, second side, Time segment 2:08–4:30. In VOAHA I, the user could quickly and immediately get to this segment by simply using the Segment Key option and entering g573.
27. Following discussion of access and control at the International Oral History Association (IOHA) in Rome, we created an opening page on the VOAHA site that requires acceptance before the user can proceed. Additionally, as noted earlier, a formal request might be required to access some materials (e.g., the furniture workers interviews).
28. Kaye Briegel was intimately tied to the program starting with her work on the petroleum project in the 1980s. Although she was not formally compensated for it, once she began teaching regularly in the History Department she served as a de facto OHRC staff member and became the co-founder of VOAHA. In 2005, she became the interim director of the Oral History Program until her retirement, when Ali Igmen was assigned to be director.

29. Despite very high ratings for the VOAHA proposal, it was initially turned down because it focused on orality and not on transcripts; and, in the second effort, an intellectual property rights lawyer on the NEH panel raised concerns. With the encouragement of the NEH Preservation and Access Director, a third proposal was submitted. Apparently, the strong responses to both of these criticisms resulted in the grant.
30. **MIT**-Accenture Digital Government Award in Higher Education.
31. Tragically, after an extended sick leave, Dave Bradley died in January 2010, and Nancy Rayner retired shortly thereafter.
32. The description of the migration process is based on communications with Jill Horn, Technology Strategist for Academic Technology Services, who worked closely with the software company, and supervised and directed the entire migration process. Any confusion about the process is due to my own limited understanding, not to her expertise.
33. Originally named Symposia, this software was developed by Innovative Interfaces Inc. It is being continually improved in response to feedback from the libraries that are using it.
34. Ali Igmen, our successor and current director of the Oral History Program, joined Kaye Briegel and me in this review and subsequent discussions with the technology strategist.
35. The Project Jukebox migration apparently has been able to keep this linkage.
36. William Schneider, Email message to author, July 2, 2012.

CHAPTER 3

Adventures in Sound: Aural History, the Digital Revolution, and the Making of "'I Can Almost See the Lights of Home': A Field Trip to Harlan County, Kentucky"

Charles Hardy III

When invited to write a chapter on "'I Can Almost See the Lights of Home': A Field Trip to Harlan County, Kentucky," the editors asked me to place it in the context of the technological changes taking place as the tremors of the digital revolution were beginning to shake and jumble oral history practice in the 1990s. What, they queried, is the backstory? What were my objectives in creating it, the practical and theoretical problems that Alessandro Portelli—upon whose interviews the essay-in-sound was constructed—and I attempted to address while working on it? And how did Gerry Zahavi—who edited this hybrid piece of experimental scholarship for *The Journal for MultiMedia History (JMMH)*—and the two of us attempt to solve those problems?

In brief, the publication of *Lights of Home* in 1999 was a collaborative effort that, for me, was the culmination of a succession of aural history projects I had been working on since the early 1980s, including oral history-based radio documentaries and audio art pieces. Very much an artifact of the 1990s, it was an experiment in digital, multiple-media publication at a time when Internet capabilities were limited. It was also part of a broader series of initiatives to convince oral historians to record higher fidelity interviews and supporting sound

Figure 3.1 "I Can Almost See the Lights of Home": Introduction.

documents and to author in sound, the same medium in which they captured their interviews. What follows is a history of its genesis, composition, and publication (Figure 3.1).

Backstory

My own work in oral history began in the late 1970s as part of a community history project in four Philadelphia neighborhoods. In the early 1980s I produced two oral history-based radio documentary series—*I Remember When: Times Gone But Not Forgotten*, on the history of working-class Philadelphians, for WHYY-FM in Philadelphia, and *Goin' North: Tales of the Great Migration*, on African American migration from the American South to Philadelphia in the early decades of the twentieth century. I continued to create radio documentaries through the 1980s until, feeling confined by the standard documentary form, I experimented in audio art.[1] At that time, analog audiocassette was the format of choice for most oral historians, and open reel remained the accepted archival media for audio preservation. The accelerating PC revolution and the introduction of the CD and CD-ROMs, however, were not only revolutionizing the entertainment industries, but also providing new media for the distribution and publication of the audio from oral history interviews. And digital audiotape (DAT), introduced in the early 1990s, provided an affordable way to record digital sound. The early tremors of the digital revolution were also raising questions about oral history practice, most directly, oral historians' choice of the equipment needed to record their interviews.

Working with oral history interviews in public radio drew me to the meetings of Oral History in the Middle Atlantic Region (OHMAR) and the Oral History Association (OHA) in the early 1980s, where folks started to turn to me for advice about field recording equipment. I was happy to oblige, for I was eager to produce radio documentaries with oral history interviews recorded by others.

In 1993, Mike Frisch, then editor of the *Oral History Review* (*OHR*), asked me to write an article on new technologies for a special issue of the journal. Soon afterward I also got a request to author a piece on "Aural History and the Digital

Revolution" for *Narrative and Memory*, and so I began a long descent down a rabbit hole. To make sense of new technologies and their potential impact on oral history practice, I needed to understand the history of *aural* history—that is, the actual recordings of spoken-word reminiscences—and sound recording and dissemination technologies.

The study quickly expanded. Finding few written sources on the history of sound documentaries that made use of aural recollections, I contacted independent radio documentary producers, archivists, old sound engineers, and radio pioneers—anyone who could help me learn more about the historical use of oral history in sound media and identify seminal works and trailblazers in aural history and sound documentary. I discovered *The March of Time* and *Cavalcade of America* sound documentaries from the 1930s; the CBS, NBC, and educational radio documentaries of the 1940s and 1950s; and producers whose works were broadcast by Pacifica, NFCB (National Federation of Community Broadcasters), NPR (National Public Radio), the BBC (British Broadcasting Corporation), and CBC (Canadian Broadcasting Corporation). I watched films by George Stoney, Barbara Kopple, and Alan Berliner, burrowed into folklore, audio art and radio theater, acoustic ecology, and the scholarly literature on sound studies, media theory, and electromagnetic sound communication. Here, beneath the surface, were these communities, unknown to each other—by and large—and sharing the same fascination with the recorded word and other sound documents.

I was soon way too deep in the tunnels to write the article for the *OHR*. When *Narrative and Memory* rejected the manuscript I sent them, I pushed forward on other projects.[2] Interest in the importance of recording high fidelity interviews and about the growing opportunities to share them on radio, CD, CD-ROM, and other media, however, continued to grow. At the 1993 Oral History Association annual meeting, I led my first workshop on sound recording and dissemination ("Thinking Sound: A Workshop on the Use of Oral Histories in Sound Presentations; Past, Present and Future"). Through meetings, workshops, and the growing number of calls I was fielding for advice on equipment, I honed my arguments to overcome a widespread technophobia among oral historians and struggled to find better ways to explain the importance of using a quality tape recorder, external microphones, and headphones, as well as learning the field-recording techniques needed to capture *broadcast quality* sound. Might this, I began to argue, also be considered a professional responsibility?

In 1992, I had the good fortune to attend the Columbia University Oral History Research Office's (OHRO) Summer Institute in Advanced Oral History Training. Here I received an extraordinarily rich crash course in oral history theory and practice and was able to bounce my ideas off of Ron Grele and Mary Marshall Clark, who invited me back in 1995 to teach a class called "Thinking Sound," on field recording and the impact of emerging technologies on oral history practice. Teaching at the summer institute also motivated me to learn who was doing the best work with oral histories in public radio and what opportunities the emerging books-on-tape industry, CD-ROM, and the Internet might offer oral historians.

In 1991, the Voyager Company had included four hours of audio when it published *Who Built America: An Electronic Book* on CD-ROM. A reviewer for *The Wall Street Journal* found the excerpts from oral history interviews to be "some of the most fascinating entries."[3] In Alaska, the team led by Will Schneider creating Project Jukebox was doing path-breaking work, not just in uploading oral histories on the web, but also explaining how authoring in multimedia alters the way one thinks about history. And in 1993, radio producer David Isay had provided two eighth-graders in a poor black neighborhood in Chicago with audiocassette recorders and microphones, and from their interviews, produced *Ghetto Life 101*, an extraordinary award-winning documentary recognized by the Corporation for Public Broadcasting.[4]

For field trips, I took summer institute participants on visits to the American Social History Project and to David Isay's studio in lower Manhattan. What are the implications for the field of oral history, we pondered, when the technology now enables middle school students, prisoners, senior citizens, and other *ordinary Americans* to become the interviewers and allows individuals with their own multi-track sound studios to produce award-winning documentaries? What impact does it have when audiences are fascinated by the spoken words of oral history interviews now accessible on the Internet, CD-ROM, and books on tape? What new wave of democratization do these developments foreshadow, and what potential changes in oral history practice?

Ron and Mary Marshall invited me back for the 1996 summer institute, and there I met Alessandro Portelli, with whom I shared an adjoining room in an otherwise empty floor of university housing. I had read and marveled at his essays in *The Death of Luigi Trastulli and Other Stories: Form and Meaning in Oral History* (State University of New York Press, 1990), so I knew of Sandro's eloquent and impassioned ideas on the "orality of oral history." Here was a kindred spirit. Moreover, he had said that he would soon be conducting more interviews in Harlan County. Why not collaborate? So one night I asked the question. Here is how Sandro remembered it:

> Charles Hardy's challenge ("You talk about oral history: how about presenting it aurally?") was, thus, an opportunity to go back to the original forms and motivations of my work, but also an occasion to step beyond those 1960s and 1970s experiments, to seek a new form of scholarly presentation in nonprint media.[5]

A Project Is Born

Our original plan was to produce a half-hour sound documentary for broadcast on public radio in the United States and abroad and to submit a jointly written article with an accompanying CD of the program to either the *Oral History Review* or *Memory and Narrative.* We tentatively agreed that Sandro and I would co-write the documentary, that I would produce it, that he would narrate it, and that Steve Rowland would engineer it on his brand new, state-of-the-art, $40,000, hardware-based Dyaxis, a Digital Audio Workstation (DAW) recently introduced by Studer. An old friend, Steve was an independent radio documentary producer whose works included the 1992 Peabody Award-winning *Miles Davis Radio Project.* OHRO would act as the institutional sponsor and permanent repository for the tapes, transcriptions, and supporting materials. Bruce Stave, then editor of the *OHR*, was excited at the prospect of publishing the first oral history study to use print and sound in a complementary fashion.

Producing the program with Steve on a bug-ridden prototype DAW rather than in an established analog, open-reel sound studio was a risk. But the Dyaxis offered unprecedented control over audio—the ability to lay each track in place and to fine-tune each fade, cross-fade, mix, and levels of multiple channels—and it replaced tape editing by razor blade and adhesive tape with nondestructive editing that could be fine-tuned within a hundredth of a second and reversed with a click of the mouse. The digital revolution was now revolutionizing and democratizing audio production as well as sound capture and dissemination.

Since Sandro would be recording his upcoming Harlan County interviews with a DAT recorder and the documentary would be distributed on CD, the Dyaxis would enable us to engineer a fully digital production. To keep down studio costs, I would purchase my own two-channel DAW to edit all of the audio files needed for mixing from my own desktop computer. When first released in 1991, ProTools cost about $6,000, and was a buggy software that I found hard to use. Since then, however, a number of companies have introduced very affordable two-track editing programs, for both MAC and PC.

Recording Interviews for the Ear

Sandro and I met again in October at the 1996 OHA annual meeting, and there, I provided him with a new SONY DAT Walkman, good omnidirectional and stereo microphones, headphones, a mic stand, and DAT cassettes. It took no more than an hour or two for Sandro to learn how to use the equipment, and for me to give a quick primer on close miking, recording ambiance at the beginning and end of each interview, controlling the sound environment, and other basic field recording techniques used in public radio.

So why a DAT recorder? In 1996 few oral historians had the training or ears to capture high-fidelity field recordings or had understood or embraced the need to so. Very few had yet made the leap to digital field recorders. DAT recorders at the time were expensive—more than $2,000 for the full-size SONY TCD 10

D-10 Pro and more than $600 for the DAT Walkman. Digital minidisc recorders were competitive with good analog cassette recorders in terms of price, but had serious drawbacks, including high compression to maximize record time. Therefore, most oral historians were still using inexpensive audiocassette recorders, often without external microphones, and they were satisfied with sound quality that was adequate for transcription. DAT, however, offered uncompressed, high fidelity stereo sound recordings that, unlike analog recordings, did not degrade—at least in theory—when duplicated.

I had been recording on analog cassettes for more than 12 years when in March 1991 I first took a SONY DAT recorder out early one cold morning on a fishing boat in the Delaware Bay. Unless one could afford a very expensive Nagra, the Rolls Royce of analog tape field recorders, rivers were tough places to record because of the high-end frequencies, wind, and engine noise. I was producing a one-hour radio documentary on the history of the Delaware River Basin Shad fishery, entitled *The Return of the Shad*, and I had struggled unsuccessfully to use river ambiance recorded with my TCD-5M (the high-end SONY audiocassette recorder) and cardioid mics. The new DAT recordings were a revelation, the sound so clean and clear that one could create digital sound documentaries of extraordinary clarity.

For our Harlan County project, I wanted Sandro to be able to record interviews and other sound documents of equal auditory clarity and power, which he could now do with the SONY DAT Walkman. Here was another opportunity to let oral historians hear the promise of the digital sound recording for their own work.

New Professional Responsibilities

Other oral historians were also responding to tremors from the digital revolution at the 1996 OHA annual meeting. Sherna Gluck was sharing her concerns about the ethical issues involved with posting oral history interviews online. Marjorie McLellan, Pamela Henson, and others, were talking about the impact of digital media on collection, archives, preservation, ethics and law, and other areas of oral history practice, and I was bending the ears of OHA president Richard Candida-Smith and president-elect Linda Shopes about the need for oral historians to record not only high-fidelity interviews, but also the other sound documents that future aural historians would need to effectively author in multimedia. There, too, the Publications Committee commissioned me to write a pamphlet on field recording for the OHA Pamphlet Series that would provide oral historians basic information about field recording equipment and its use. It would also include a CD with excerpts of interviews recorded with different equipment under different conditions, to help those who used the pamphlet develop their ears and better record broadcast-quality interviews.

In August 1997, Richard charged a new Technology Update Committee "to investigate the use of oral histories in various nonprint media that are sometimes referred to as 'new technologies'" and to make recommendations for revision of the OHA *Evaluation Guidelines.* Chaired by Sherna Gluck, the committee included Terry Birdwhistell, Pamela Henson, Marjorie McLellan, Roy Rosenzweig, and me.[6]

The Harlan County project dovetailed perfectly with both of these initiatives. Our sound documentary, if well done and well received, could demonstrate what oral historians could produce with good equipment and high-fidelity recordings and strengthen proposed revisions to *Evaluation Guidelines*—revisions that I hoped would include a clause that oral historians should treat the sound recording as a primary source of equal value to the edited transcript (see Appendix A). Our sound documentary and the field-recording pamphlet would join the voices of other oral historians doing innovative work in archives, programming, and elsewhere, and accelerate the arrival of the paradigm shift in oral history practice that I believed was soon to take place. The historical precedents for this were clear in the history of other technological shifts: film as the one-camera carrier of live theater performances before the development of cinematography; the steady beat and simple melodies of music in the age of sheet music superseded by improvisational jazz in the age of the phonograph and the subsequent development of the art of phonography, which enabled records to challenge live performances as the *real* music event.[7]

While Columbia transcribed Sandro's interviews and the committee began to think about its charge, I pushed ahead with the field-recording pamphlet. I planned to expand the focus from a basic how-to manual for monaural audio recording to a broader introduction that argued for the recording of high-fidelity interviews in two-track mono, stereo, and on videotape by talking about emerging technologies and future digital uses of oral history interviews and supporting soundscape and sound event recordings.

A Script Is Born

After Sandro completed his Harlan County interviews that October, he sent the DATs to me. I then sent audiocassette dubs to Columbia, where Mary Marshall had the more than 20 hours of audio converted into 700 pages of transcription. At some point, she and Ron decided to integrate the development of a script for our Harlan County documentary into the next year's summer institute. As they arrived by mail, I pored through the transcriptions and struggled to find the pivotal events and chronology needed to block out a documentary.

As the summer institute approached, I was deeply worried about how to shape the documentary. That June, the night before the first day of the institute, Sandro arrived from Italy and we began to discuss what we might do. At some point, we decided to record our conversation, so I set up my DAT recorder with two mics and pressed record. Here is how that interview, and how *The Lights of Home* both began.

> ***Hardy***: Ah, there we go. Yes. I see it moving. Okay.
>
> ***Portelli***: Are you ready?
>
> ***Hardy***: Yes. I should be on mic two. We've got our two-track mono going here. I'll move that down a little bit. Let's start from the beginning. Last October, you went back to Harlan County, and you did a series of new interviews. What were you trying to get?
>
> ***Portelli***: Well, basically I was trying to get some decent-sounding tape and try to experiment on using another medium to present my findings. I began fieldwork in Harlan County in '86. So that was ten years. Also, for the first time, I was getting a chance to air some of my views in front of people in the community. Because there was a seminar that the University of Kentucky was setting up in Benham, which is one of the mining towns in Harlan, and I got a chance to speak to them.

As I asked my questions and Sandro shared his thoughts about the interviews he had conducted the previous summer, the transcripts, for the first time, began to come to life and make sense. At last, I began to see the order and the patterns of Sandro's mind and saw a way to give the material shape.[8]

The story began to come to life as Sandro explained the subjects and themes of his interviews. The interview also had an impact on Sandro, as he later explained in the *JMMH*.

> In sound, we could include interpretation and analysis on the same level as the voice of the narrators, and yet (thanks also to my non-native English) distinct from them. The solution we found was for Charles to interview me about my thoughts and experience. This created a continuity between our dialogue and the dialogic interviews in the field. It created a space in which I could articulate my hypotheses and conclusions, but make them tentative, dialogic, imbued with my own subjectivity and history as a corrective to the impersonal authoritarian attitude of the scholar analyzing his data. I was not even in total control of the agenda. When Charles asked me, unexpectedly, "Why are you so concerned with death?" he forced me to look at the relationship between my times in Harlan and my life in Rome in ways that I had not articulated before, and to understand them both (and myself) better. It may not be a coincidence that, after that interview I went back to Rome and started working on an oral history of the Nazi massacre at the Fosse Ardeatine in Rome during World War II.[9]

Over the next two weeks, the documentary quickly took shape. Each evening Sandro and I worked late into the night sharing our favorite actualities (audio clips), looking for connections, and talking through the themes and the stories that he wanted to tell. As we pored through the binders of transcribed interviews, I cut and pasted actualities on my laptop and banged out a rough draft to hand to Mary Marshall for photocopying and distribution. On Tuesday, June 17, institute participants helped us analyze the tapes and transcripts. By Saturday, we had the first draft of a completed script, which Mary Marshall photocopied and handed out for group review the following Monday, after which we hammered out a second draft, which was photocopied, distributed, and then discussed on the last day of the institute.

In these work sessions, the institute participants pushed us to explain and better make sense of our objectives and decisions. Voicing, as Sandro explained above, was a major conceptual challenge. Sandro did not want to narrate the documentary, but the material made little sense to the institute participants and me without the explanation and context that he provided us. An alternative, however, proved close at hand. Why not use excerpts from our interview recorded the night before the summer institute began? Mary Marshall quickly had the audio transcribed, and interview segments, as we had hoped, pulled the pieces together in a new and exciting way. What, after all, could make more sense than to interweave the stories of the people contained in Sandro's interviews with the story of Sandro and me attempting to make sense of those stories? Conversations about conversations reflected Sandro's ideas about oral history as "an experiment in equality" and my ideas about new forms of aural discourse. Listeners could now eavesdrop on, and share in the process of interpretation, of making sense of it all. Listeners could now engage with Sandro and me in the search for meaning. Here is how Sandro described it in his article in the *JMMH*.

> The other, important limitation of these projects was that one tended to be carried away by the documentary impulse. The possibility of presenting documents first-hand, rather than just writing about them, was accompanied by the unspoken persuasion that they were all but self-explanatory. This problem seems to persist in most video productions in oral history, where both the deceptively objective (and unexplained) documentary montage and the banal, often boring explanatory "talking head" (often, in mere transposition of written discourse) are equally authoritarian, especially when accompanied by an esthetics and rhetoric of discourse dominated by the grammar of television documentary aimed at an impressionistic, quick fruition. Most importantly, by suppressing the researcher's presence in the field (either by eliminating her altogether from the final product, or by separating her commentary from the documents), these modes of presentation frustrate the most original contribution of oral history to the practice of field work and to the forms of presentation of field material: the dialogic approach, the interpersonal encounter between researcher and narrators, from which both emerge with a new, different awareness.[10]

By the end of the summer institute, our piece also had a name, which had emerged in our interview and which would open the "Fifth Movement: Recapitulation and Prologue."

Song: *"I Can Almost See the Lights"*

Hardy: Okay. Let's talk about some of the tape that we're going to be working with. You did interviews with a half dozen different people. Can you tell me about the one or two moments that absolutely stand out, that you had one of those moments of epiphany and you said, "This I'm going to use."

Portelli: Okay, one was a story of seeing the light of the city. It's not final, but at this point I see it as a title for the whole project. "I Can Almost See the Light of the City."

(Hiram singing on right channel)

Hardy: Is this "I Can Almost See the Lights of Home"?

Portelli: The song is "I Can Almost See the Lights of Home." And then the story is about the city. "I Can Almost See the Lights of Home." Yes. So I can think of it as a title because it's both otherworldly, but then very concrete, and has to do with, of course, this theme of death and the road and symbolism. So that was one moment.

And then, of course, the interview with Dee Dee—Annie's granddaughter...[11]

By June 27 we had a 28-page script for a program of closer to three hours than the half hour we had first envisioned. The implications were clear. The documentary, which was no longer a documentary in the traditional sense, had become a multi-part series, far more expensive to produce and far harder, if not impossible, to publish in a journal because of its length, which would require explanatory text and multiple CDs. It would now also be harder to market to public radio stations that preferred programs that fit into one or more neat half-hour or one-hour blocks of time. Broadcast of *Lights of Home* would require major reworking of the script to break it into uniform, artificial, 30-minute segments.

From Type to Tape...or Sound Files?

The summer institute now over, the next step was to convert the words on the page into sounds in the ear. What reads well does not necessarily listen well, so I needed to audition the actualities we had chosen and hear how they sounded when assembled together. I also had to select the music, ambiances, and sound effects needed for punctuation and context—to locate the interviews in the soundscapes of their origin, to provide auditory description, and to amplify, deepen, and add meaning to the interpretation. In the past 15 years, I had produced sound documentaries working both from audio without the benefit of transcripts and from printed transcripts, cutting and assembling the actualities for a program conceived on paper. Working from a transcript is a far more efficient way to produce a long-form documentary, but what works well for the eye does not always work for the ear. Conversely, the ear will at times find unexpected riches in what makes little sense to the eye.

Back in West Chester, I had to figure out how to edit the audio files. Fortunately, Macromedia had recently introduced SoundEdit 16 and Deck II. A software-only recording studio less expensive and easier to use than ProTools, it had 16-bit resolution; sample rates of 24, 44.1, or 48kHz; up to 32 tracks; visual, nondestructive editing; a virtual mix board with knobs for panning and faders for levels; a decent digital equalizer; some simple effects programs; compression for final mixing; and other features.

Deck II also provided me the ability to engineer *Lights of Home* myself on my new Power Mac. This meant that production would not remain only digital. I would have to import audio from the DAT recorder into the Power Mac through a stereo-analog audio mini-plug. The preamps, however, were quite good, and the sound, surprisingly clean. The DECK II sound files, however, ate RAM (Random Access Memory), so I had to export and save completed segments onto seven large, external hard drives, each of which cost about $100. (Today *Lights of Home* can easily fit onto a single, small, inexpensive thumb drive.)

Deck II proved remarkably easy to learn to use and was bug free. Up and running within a few days, I lost few files over the months that I worked on the project. Untethered from an engineer and with studio costs billed by the hour, I was free to experiment with timing, mixes, and transitions, to take out and restore phrases, and to compose my own movements of *contrapuntal radio*—dense, multivocal montages based upon a musical rather than a linear, typographic model.

An Aural History Essay in Sound

Having produced sound documentaries and audio art pieces for public radio and live performance from 1980 to 1992, I was acutely aware of how hard it would be to convince NPR-affiliated public radio stations to broadcast a piece as uncommon as *Lights of Home*. After a period of experimentation and looser formats in the 1970s, public radio during the 1980s had solidified into more rigid half-hour and hour programming blocks, populated by highly produced programs coming out of NPR and APR (American Public Radio), and a flood of programming produced by affiliate stations and independent producers.

Sandro's free form of interviewing precluded a traditional documentary approach. The tape lacked the clear references to historical events needed to produce an actuality-based sound documentary. And, I had already decided to produce *Lights of Home* more for CD distribution than radio broadcast. If *Lights of Home* was to demonstrate the possibilities and advantages of authoring in sound and multiple media, the piece needed to emerge organically. I needed to listen to the sound files and let them tell me how to give them life and shape, rather than try to box *Lights of Home* into an existing genre, formula, or format.

Freed by DECK II from studio costs, I could approach *Lights of Home* as a work of audio art and integrate all that I had learned in the 1990s from listening to Glenn Gould, Dmae Roberts, and artists working with spoken-word recordings. Here was the opportunity to "think in sound" in a way that would win the ears and hearts and minds of oral/aural historians—to demonstrate how sound could carry serious scholarship and could, in fact, capture nuance and meaning and aspects of human experience and history muted by the printed word.

We had solved the problem of voicing at Columbia, but not how to structure such a long piece freed from the conventions of radio broadcast and sound documentary. How could I retain the complexity, mystery, and contradictions at the heart of these people's lives and Sandro's own world-view and scholarship? Of the Italian communist immersed in American literature and folk song trying to find out what he could learn about America and Annie and Chester Napier and his other narrators and himself by asking questions and listening? What were the synchronicities? Where did they all resonate together?

I also had to grapple with temporality—that is, with how to layer and unlayer moments in time. In the interviews, Sandro's narrators mixed past and present, shared family and folk histories spoken to them by parents and grandparents and old-timers, and they were now passing those along to Sandro. There were so many layers. Sandro had been interviewing in Harlan since 1983, at times asking the same people the same and overlapping questions, so there were instances where they were revisiting topics that they had previously explored.

From responses to my radio documentary series in the 1980s, I knew how many listeners found competing voices confusing and loud musical beds deeply annoying.[12] But I had also spent many hours listening to and thinking about the best sound works using aural reminiscences produced over the past half century, and how effective communication in sound required listeners to reprogram their expectations and their ears. I disliked the radio news formula of information immediately comprehensible in a single listening. Listeners should also learn to slow down, savor, give up the need for control, and approach a sound essay in the same way they listened to a new piece of music, read a novel or poetry, or viewed a painting. They should treat the sound essay as something to be savored, as something that might perplex at first listening, but that drew them to listen again. Something that would improve, like Glenn Gould's *Solitude Trilogy*, rather than bore with repeated listenings. History is not simple or linear or easy to understand. It is complex and mysterious, contradictory and multivocal. And this was something that I experienced more deeply when conducting oral history interviews than I did through my formal historical training in graduate school or my work with written primary sources. Historical study requires the historian—amateur or professional—to grapple with those complexities. Shouldn't aural history convey that same complexity—how patterns and meaning emerge out of contending voices, how cosmos emerges out of chaos—rather than provide listeners with predigested, reductionist, oversimplified redactions that provide us the illusion of understanding something that in fact we do not know?[13]

Freed from the need to please the overworked gatekeepers at public radio stations, I could produce an essay to demonstrate the opportunities of this format not just to oral historians, but also to scholars in other disciplines, books-on-tape listeners, college students, opera lovers, and others who engaged in the beauties and mysteries of life through their ears. I wanted all of these groups to think about Hillbillies and communists and the people of Harlan County and organic intellectuals, American history and contemporary America, and Americans' attitudes toward and relationships with foreign countries, and to consider historical scholarship, aural history, applied media theory, and electromagnetic communication in new ways.

To keep the project manageable, Sandro and I decided to limit *Lights of Home* to the digital, stereo, high-fidelity interviews he recorded in October 1996. This also gave us our subtitle: "A Field Trip to Harlan County, Kentucky." To make sense of these, however, we also needed our two-hour interview from June 1997. And we added one more piece of tape: an interview with Mildred Shackelford, a miner, political activist, and poet that Sandro recorded on audiocassette on November 2, 1990. This would produce one of *Lights of Home's* most memorable moments. In "Chapter 3: The Third World Suite," one can hear Sandro, on the left channel, restating Mildred's words in our June 1996 interview, and Mildred's voice, on the right channel voice, speaking almost the exact same words in 1990.[14]

Aural history as lived only occupies the moment it is spoken. Once recorded, and now captured in space, it can be reanimated in new temporal moments and contexts. So, to me, a traditional historical approach made no sense for *Lights of Home*. Way too reductionist, it would suck the life out of the interviews and the sound piece. How does recorded sound enable one to explore and make sense of and feel contradiction and multivocality in different ways—one might argue better ways—than the linear printed word?

The first task was to listen to all of the actualities and to block out program segments. Here I soon heard that some parts of our script needed explanation and contextualization while others were most compelling when spoken or sung without comment. I soon also became acutely aware that I did not have enough sound to produce *Lights of Home*. I would need to go to Harlan to record the music, sound effects, chatter, church services, greetings, opening and slamming doors, dogs barking and crickets chirping, background ambiance—the full range of soundscapes, sound markers, and sound events needed to bring these people's stories and worlds to life. So Sandro agreed to return to Cranks Creek, with me in tow, after the 1997 OHA annual meeting in New Orleans that coming October.

At that OHA meeting, the New Technology Committee met for the first time as a group, and I gave an update on the field recording and equipment manual to the Publications Committee. That Sunday I flew to Harlan and then spent two days recording with Sandro. There we digitally recorded more than two hours of music, a storefront church service, family picnic, soundscapes around the Napier and Gent homes, roosters and chickens, a thunderstorm and rain beating on tin roofs, conversations in automobiles and living rooms, and more.

With all of the sound elements now in hand, I went back to the DAW. Producing *Lights of Home* was a juggling act in which I attempted to keep all the different stories, theory, narratives, emotions, politics, voices, and sounds in motion together, understanding from the start that every listener would hear, experience, and make sense of it differently. I would consider myself successful if I managed to intrigue a few people enough so that they would listen the second and third time, necessary to both hear the material with new ears and to really enter the interpenetrating worlds of Sandro and his friends and acquaintances in Cranks Creek.

As structured by the 28-page script, *Lights of Home* was a very long piece with lots of time to include a broad range of narratives, montages, and musical interludes. In production, certain segments gravitated toward history and current events, while others worked best when focused on narrators' interior lives. The latter worked best when the mixes were more impressionistic than narrative, and for me, these required more attention, creativity, and experimentation. The former I came to label "chapters," and the latter "movements" (Figure 3.2).

One of Sandro's insights about Harlan County was the simultaneous coexistence of extraordinary violence and extraordinary care for one another. In the script, this was expressed most compellingly, in two conversations with Chester and Annie Napier. Sandro had recorded both of them recounting a horrific accident that Chester had suffered while driving a coal truck. Even after repeated editing, sequencing, and conservatively intercutting the audio of their separate

Figure 3.2 "I Can Almost See the Lights of Home": Second Movement.

narratives, however, each actuality felt too long and was missing something essential about the accident and their relationship. The contrapuntal, polyphonic layering of the two vocal passages—of two different vocal instruments, each playing their own variation on the left and right channels—brought Sandro's vision of this simultaneity to life. It took many hours of experimentation to work out the timing, pulse, levels, and cross fades in a close to ten-minute polyphonic montage titled "Violence and Care." How long could the two voices play together on separate channels, one slightly more audible, before the next voice needed to swell into foreground? How many seconds could I sustain the two voices at equal volume and the equality of the two conflicting narratives? The passages needed to be long enough to enable a listener's ear to gravitate to one voice or the other but short enough for the cacophony to not drive audiences away before the mix moved them to the phrase of words that kept the back and forth of violence and care in dynamic tension.

This was just one of the compositional concerns one faces working in sound media—an aspect of thinking in sound that authors in digital sound would need to master if they aspired to do more than mono-vocal sound narratives. One can only do so much writing about sound studies. At some point, sound studies scholars would need to start authoring in sound as well, and to do this, they would need examples compelling enough to motivate them to exert the time and effort to learn new ways of listening, thinking, and communicating.

This, of course, was exactly what the digital revolution was requiring us to do. Multichannel sound was more than a novelty or gimmick. The movement from mono to stereo to multichannel and binaural sound could draw us back toward the full three-dimensionality of natural sound communication. It would require the relearning of old ways of listening, what R. Murray Schafer called "sonological competence," that had withered when the printed word reshaped culture and education in the age of typography.[15] An analogy that I had begun to use in workshops was to stereo vision and the autostereogram, in which one must overcome normal vergence—a two-dimensional image—to see an otherwise hidden three-dimensional image, visible only when one's eyes unfocused.

To let listeners *hear* the dynamic simultaneity of human experience masked by writing, something Gould had pioneered in his *Solitude Trilogy*, I produced two polyphonic "movements" in *Lights of Home*, neither of which was as ambitious or challenging to the listener as passages created by Gould. I hoped, however, that attentive listeners would experience the same sort of epiphanies that had drawn me to this form of expression.

The Journal for MultiMedia History

At 1998's annual OHA meeting in Buffalo, Sandro and I presented a session on "'I Can Almost See the Lights of Home': The Making of an Oral History Essay in Sound." There, we played rough mixes of one of the chapters and one of the movements, talked about the project, and shared our thoughts about the challenges of finding a publisher and broadcast outlets. In the audience was Gerry Zahavi, who, along with co-editor Julien Zelizer, was about to publish the first issue of *The Journal for MultiMedia History*, a new, experimental online journal devoted to "re-conceptualizing the craft and art of historical research and pedagogy." "The *JMMH* is," the editors noted in the journal's founding statement, "the first peer-reviewed electronic journal that presents, evaluates, and disseminates multimedia historical scholarship."[16]

After the session, the three of us met, and Gerry asked if we would consider publishing in the *JMMH*. It did not take long to accept his offer. The journal would post as much audio as we wanted, and we would have no space limitations on accompanying print materials. *Lights of Home*, too, would now be published in a journal whose objectives were in complete harmony with our own.

In the months that followed, Gerry and I brainstormed over what to call this multiple media publication and how to organize its various components for web audiences that could potentially include "an entire universe of interested readers."[17] Web publication also raised some very fundamental questions about word choice. What exactly was it that the *JMMH* would be publishing? An article? A documentary? A history? Back in 1976 Glenn Gould had called "The Way of North" an "oral tone poem."[18] Dmae Roberts used the word "docuplay" when describing her 1989 Peabody Award-winning "Mei Mei, A Daughter's Song," which was an oral history-based exploration of her mother's early life in Taiwan and Roberts' relationship with her mother.[19] Neither of these, however, made sense for our project. After some discussion, it dawned upon me that *Lights of Home* might best be described as a *sound essay*, or better yet, an *essay in sound*.

This, then, raised another question. One writes an article, produces a radio documentary, directs a film or video documentary, and composes a piece of music, but what were we doing? Since *writing* was too bound to the written word, we decided to distinguish between the verbs *author* and *write*, using the former as the default and the latter only when the act of composition was specific to the written word.

How, then, should we refer to those who listened to the digital audio files and read the printed script and essays written by Sandro and myself in this multiple media publication? *Readers* clearly made no sense, and *users* was too closely associated with drug use. So we settled upon *browsers* which, as natural as it sounds today, felt quite inapt at the time. A browser, after all, was a software application, not a person.

A simpler but also important need was to clearly distinguish *oral* from *aural* history. Although British and Canadian scholars had long embraced a distinction between the *oral* history on the page and the *aural* history in the ear, American

Figure 3.3 "I Can Almost See the Lights of Home": Contents.

oral historians in the 1990s were less aware of the profound differences between the two (Figure 3.3).

In his editor's introduction, Gerry did a wonderful job summarizing our desire to create a new mode of thinking about and presenting oral history:

> a new aural history genre that counterpoises the voices of subject and scholar in dialogue—not merely the dialogue that takes place in the real time of an oral interview, but the one that occurs as interpretations are created and scholarship is generated... an instructional manual on authoring in sound and a manifesto of sorts. It challenges oral historians to truly explore the full dimension of the sources they create and utilize in scholarship—to engage the "orality" of oral sources. It challenges all historians to consider alternative modes of presenting interpretations, modes that render the very act of interpretation more visible while preserving and respecting the integrity of primary sources.[20]

When finally published in volume 2 of the *JMMH* in 1999, *Lights of Home* included an editor's introduction; a table of contents ("An Essay in Sound") that hotlinked to 17 separate audio files, downloadable or streamed at 28.8K, 56K, or a blazing 80K; the "Script," which included a full transcription of the essay and photographs; Sandro's "Field Notes from Harlan County," a 2,800-word history of the project and his own interest in sound; and my 9,000-word article on "Making an Essay-in-Sound." Prompted by Gerry to make it easier to read on the web, I structured the article as a series of short *probes* on narrative voice, the history of the sound documentary, and the impact of new media upon forms of discourse, as well as the composition, structure, and editing of *Lights of Home.* Publication on the *JMMH* enabled me to hotlink to audio clips from three *The March of Time* programs from the 1930s and early 1940s, to chapters and movements referenced in the essay, and to the rough mix of a chapter that I scrapped because it did not sound as well as it read on paper[21] (Image 4: Screen shot, Table of Contents).

While Gerry, with the help of Susan McCormick, worked on the programming and layout, I continued to produce the essay in sound, sending CDs of produced segments to Sandro for review. When it was near completion, I mailed CDs of the program to Annie Napier in Harlan County for her to review and share with other narrators. Their only suggestion was that I include more music. Last tasks included the selection of someone to voice the host narration, since the presence of my voice in the body of *Lights of Home* meant I should not also play that role. Danish Holocaust scholar Stig Hornshoj-Moller, who was then visiting West Chester, filled this role. His voice added another wonderful tonal quality to the production and further internationalized the sound—a Dane introducing a piece about an Italian and his relationship with Americans.

The *JMMH* provided a tremendous opportunity to explore web publication. Gerry's architecture was terrific, the webpages looked great, and with a good computer and high-speed connection, the audio sounded quite good. Web publication in 1999, however, had its limitations. I produced *Lights of Home* as a multilayered, stereo, high-fidelity sound work, designed for listening through headphones or good speakers, the same way that an audiophile listened to a good piece of classical music or a long-form radio documentary. At more than two and a half hours, *Lights of Home* was much too large to stream or download, so Gerry uploaded the audio in 17 separate files. After listening to a single module, browsers then had to go back to the Table of Contents and click on the next hotlink. "It is difficult to close one's eyes and lose oneself in the world of Harlan County, Kentucky," noted a George Mason University student in 2004, "if one must regularly stop to wrestle with problems resulting from modem speed too slow to smoothly support RealPlayer."[22]

Wrestling with modem speed was not the only problem. Listening to monaural sound files, browsers could not hear the auditory space and movement that brings the voices and soundscapes to life. Flat sound also made it much more difficult to hear the audio to separate, follow, and feel the movement of the voices and sounds in "Violence and Care" and the other complex polyphonic movements. Stated more simply, *Lights of Home* listened to in mono sounds flat and one-dimensional; one listens at it rather than in it.

Another George Mason student who critiqued *Lights of Home* as an assignment in Rob Townsend's HIS 615: Clio Wired course in fall 2001, explained this quite compellingly.

> The main problem with the "I Can Almost See the Lights of Home" is that the invitation, "to close the eyes and listen to the essay" does not work on a computer. It doesn't work, because the sound quality of most speakers in computers is so terrible, that it is very hard to listen to Charles Hardy's sophisticated sound mix. He sometimes mixes interviews, songs, and aural impressions in one piece—a collage like this would sound wonderful even from a simple radio or simple tape recorder, but is too much for a computer. As long as the sound of computer loudspeakers is tinny as it is, sound pieces should be simple and

> short—focusing on an interview, or a song, or just one noise at a time—sound collages just sound shallow from those small computer loudspeakers. In addition, browsers are different from radio and tape recorder listeners—even if they are the same person. A radio is light, and mobile—one can listen to the radio in the car, or at home while chopping garlic or taking a shower. It is possible to listen to long pieces, while doing something else on the side, listening at times more, at other times less attentive. A computer, in contrast, is bulky, it needs to be plugged in not only to a plug, but also to the telephone cable, preferably indoors. A computer requires concentration and attention—people don't expect to relax and close their eyes in front of a computer, but they want to look, browse, communicate back and forth and to generally engage some action. Even though the intention of the essay is good, it did not work in the digital form. The sound pieces are too long, the sound is too complex, and it is hard to follow the narrative for the listening browser.[23]

Another flaw that became distressingly apparent as Gerry prepared *Lights of Home* for publication was my own media myopia. One of Gerry's first requests was for photographs. Neither Sandro nor I, however, had brought a camera to Harlan, or asked to borrow photographs from his interviewees for duplication. This failure has gnawed at my conscience ever since and deserves some explanation here for what it tells about media in the 1990s.

In 1996, 35-millimeter film was still the industry standard. Because of the high costs of professional printing and inclusion of photographs in print publications and my focus on radio broadcast, I had pretty much given up on photo documentation in the 1980s. In 1996 and 1997, when Sandro and I recorded in Harlan County, the affordable, digital, still cameras then being introduced into the consumer market captured only lower quality images. Nor did I know how inexpensive and easy it was to mount still images on the web. The decision, then, was to place the limited budget I had into the best digital field recording and production equipment that I could afford.

Gerry was able obtain a few images from The Appalachian Archive at Southeast Community College (University of Kentucky).[24] Sandro provided some additional photos from Annie Napier to include with the first page of the script. The paucity of photographs, however, remains to this day a source of tremendous personal embarrassment and a humbling reminder of the need to follow one's own advice. For years, I had been advising students and workshop attendees to take good photos of their oral history interviewees.

Reception

Initial responses to *Lights of Home* after publication in the *JMMH* were gratifying. Despite its many flaws, it won the 1999 Oral History Association's biennial

Nonprint Media Award and was hailed by some as a pioneering work of digital scholarship. Roy Rosenzweig, founder and director of the Center for History and New Media at George Mason University, called it "a brilliant melding of form and content and probably the best use of the Web for scholarship that I have seen."[25] It was, and still is, used in college courses on oral history, documentary across media, web genres in English studies, digital history, and media studies.

As noted above, however, student reviews posted online exposed some of the limitations of web publication in 1999 and the broad range of responses that browsers have had to *Lights of Home*. Discussions with students assigned *Lights of Home* in an upper-level historical methods course I taught at West Chester University in the early 2000s provided me with additional insight into student experiences of the essay. Some had difficulty with the Kentucky accents. Listening on high-fidelity stereo CDs, however, most were excited at how the voices and sounds brought these people and their stories to life. More gratifying was their appreciation for listening to Sandro's and my conversations about the interviews and also listening to the interview excerpts that we were discussing and attempting to better make sense of enabled them to understand how scholarship takes shape, and in doing so empowered them to reach their own conclusions about the meaning and significance of the words shared by Sandro's narrators.

Publication of *Lights of Home* on the *JMMH* freed me from having to shape it for radio. I did, however, send out CDs to radio documentary producers, book editors, and public radio station programmers for comments and suggestions about dissemination. Radio carriage was limited but did include Appalshop's WMMT in Whitesburg, Kentucky, and WRPI, which broadcasts *Talking History*, in Troy, New York.

Conclusion

Digital technologies and media have come a long way since publication of *Lights of Home* in 1999, and have since then transformed oral history practice. The impact of the digital revolution has in ways been far greater and faster than I imagined, and in other ways slower. The greatest changes may well have taken place in oral history preservation and curation. Analog tape technology has all but disappeared, and open-reel audiotape, once the only acceptable audio preservation media, is now obsolete. Indexing has been replaced by metadata. And who, in 1998, would have imagined that less than 15 years later, leaders in the field would be arguing against the transcription of oral history interviews? Growing numbers of oral historians have embraced videotaping oral history interviews and lost their fear of a medium most in the 1990s considered far too intrusive. At the 2012 OHA annual meeting, the editors of the *Oral History Review* hosted a session to discuss web-only publication of the journal. Today, people can record oral history interviews on their cell phones, edit them on laptops and portable devices, and post them on YouTube or Twitter.[26]

With the democratization of video, however, the eye continues to assert its dominance and consign the ear to a supporting role. The movement from oral to aural history has been much slower than I had hoped. Few oral historians have discovered the rewards and pleasures of authorship in sound that moves beyond simple editing and recombination of audio files. Training in sound authorship and sonological competence remains rare, even in the growing academic subfield of sound studies. *Lights of Home* remains an anomaly, perhaps still ahead of its time or forever on the margins. It can, however, now be heard as originally produced, in high-fidelity stereo. In 2014, the *JMMH* republished the audio files for *Lights of Home* as stereo MP3 files for both individual chapters and movements, and as two large files, as originally intended, for continuous listening. I hope that, when listened to at these levels, our essay in sound still offers some lessons worth learning and some experiences worth hearing.

Notes

1. To learn more about these projects, see Charles Hardy III, "A People's History of Industrial Philadelphia: Reflections on Community Oral History and the Uses of the Past," *Oral History Review*, 33(1) (2006): 1–32; and Charles Hardy III, "Painting in Sound: Aural History and Audio Art," in *Oral History: The Challenges of Dialogue*, ed. Marta Kurkowska-Budzan and Krysztof Zamorski (Amsterdam: John Benjamins Publishing, 2009), 147–167. Programs from *I Remember When: Times Gone But Not Forgotten* (WHYY-FM in Philadelphia, 1982–1983) and *Goin' North: Tales of the Great Migration*, (oral history interviews on deposit at the Atwater Kent Museum, 1984–1985) can be heard at http://www.talkinghistory.org/hardy.html.
2. In 1994 I began work as the principal project historian, lead writer, and editor of *United States History Video Collection* (Library Video Company, Schlessinger Video Productions, 1996), a stand-alone video history textbook. In 1999 Gerry Zahavi would mount the completed study, "Authoring in Sound: An Eccentric Essay on Aural History, Radio, and Media Convergence," online. In 2005, two revised sections of the study were published as the following: Charles Hardy III and Pamela Dean, "Oral History in Sound and Moving Image Documentaries," in *Handbook of Oral History*, ed. Thomas Charlton, Lois Myers, and Rebecca Sharpless (Walnut Creek, CA: AltaMira Press, 2006), 510–562; and Charles Hardy III, "Authoring in Sound: Aural History, Radio, and the Digital Revolution," in *The Oral History Reader*, 2nd edition, ed. Rob Perks and Alistair Thomson (London: Routledge, 2006), 393–406.
3. Walter S. Mossberg, "'Who Built America' Reveals Real Potential Of Electronic Learning," *Wall Street Journal*, September 2, 1993.
4. Hardy, "Authoring in Sound," 395.
5. Alessandro Portelli, "Field Notes from Harlan County, Kentucky," "'I Can Almost See the Lights of Home': A Field Trip to Harlan County, Kentucky," *The Journal*

for MultiMedia History, 2 (1999), http://www.albany.edu/jmmh/vol2no1/lightsportelli.html.

6. Richard Candida Smith, email to Oral History Association Committee on New Technologies, August 21, 1997.
7. On the development of the art of phonography, see Evan Eisenberg, *The Recording Angel: Music, Records and Culture from Aristotle to Zappa* (New York: McGraw-Hill, 1987).
8. Charles Hardy III and Alessandro Portelli, "Script," "'I Can Almost See the Lights of Home': A Field Trip to Harlan County, Kentucky," *The Journal for Multimedia History*, 2 (1999), http://www.albany.edu/jmmh/vol2no1/lightsscript.html.
9. Portelli, "Field Notes from Harlan County, Kentucky," http://www.albany.edu/jmmh/vol2no1/lightsportelli.html.
10. Ibid.
11. Hardy and Portelli, "Script," 1999.
12. Studying the impact of musical training upon the nervous system, scientists at Northwestern University found that people with musical training have enhanced cognitive and sensory abilities that enable them to hear speech in challenging listening environments better than those without musical training. See Nina Kraus and Bharath Chandrasekaran, "Music Training for the Development of Auditory Skills," *Nature Reviews Neuroscience*, 11 (August 2010): 599–605. The musical training or lack of musical training of both aural historians and their audiences, then, may have a significant impact on the sound of future aural history scholarship and digital publication. For more on listener responses to *I Remember When*, see Hardy, "A People's History," 21–24.
13. For more on Glenn Gould's *Solitude Trilogy* and the influences that shaped my approach to production of *Lights of Home*, see Charles Hardy III, "Making an Essay in Sound," *The Journal for Multimedia History*, 2 (1999), http://www.albany.edu/jmmh/vol2no1/lightshardy.html.
14. Sandro would use this interview segment in the first chapter of *They Say in Harlan County, Kentucky: An Oral History* (New York: Oxford University Press, 2011), 7.
15. I still consider R. Murray Schafer's *The Tuning of the World* (New York: Random House, 1977), rereleased as *The Soundscape* in 1993,—the best introduction to the history and systems of sound communication, and an essential reading for anyone interested in sound media and aural history. His World Soundscape Project, based at Simon Fraser University, is often credited for initiating the modern study of acoustic ecology.
16. See Gerry Zahavi and Julian Zelizer, "Founding Statement: About the *Journal for MultiMedia History*," *The Journal for MultiMedia History*, 1(1) (Fall 1998), http://www.albany.edu/jmmh/vol1no1/introduction1.html.
17. Ibid.
18. Glenn Gould, *Glenn Gould's Solitude Trilogy: Three Sound Documentaries*. CBC Records (Audio CD), liner notes, 1992.
19. Hardy, "Authoring in Sound," 400.
20. Gerald Zahavi, "Editor's Introduction," "I Can Almost See the Lights of Home," *The Journal for MultiMedia History*, 2 (1999), http://www.albany.edu/jmmh/vol2no1/lights.html.

21. Charles Hardy III, "Making an Essay in Sound," 1999, http://www.albany.edu/jmmh/vol2no1/lightshardy.html.
22. "The Promise of Digital Scholarship," http://mason.gmu.edu/~jcox/four.htm, accessed June 18, 2004 (site discontinued).
23. These comments came from a journal entry critique and comparison of "Lights of Home" and "City Sites" http://mason.gmu.edU/~khering/scholarship.html, accessed January 28, 2002 (site discontinued).
24. Today's Southeast Kentucky Community and Technical College.
25. Roy Rosenzweig, email to Charles Hardy, April 4, 2001, quoted in *Digital Scholarship in the Tenure, Promotion, and Review Process*, ed. Deborah Lines Anderson (Armonk, NY: M.E. Sharpe, 2004), 114.
26. For an excellent introduction to these transformations, browse the essays on collecting, curating, and disseminating in *Oral History in the Digital Age* (Washington, DC: Institute of Museum and Library Services), http://ohda.matrix.msu.edu/.

CHAPTER 4

"I Just Want to Click on It to Listen": Oral History Archives, Orality, and Usability

Douglas A. Boyd

In 1978, Paul Thompson stated that "the tape recorder not only allows history to be taken down in spoken words but also presented through them...The words may be idiosyncratically phrased, but all the more expressive for that. They breathe life into history."[1] Like many others, I was drawn to oral history by the power, emotion, and the content conveyed in the recorded voice, the expression of first hand memories tangled up in the engagement of an interview, culminating as recorded narrative performance. Folklorist Kenneth S. Goldstein began the influential book *A Guide for Field Workers in Folklore* with the reminder, "The basis of any scholarly discipline is the materials with which it deals. Without such materials there can be no subject for scholarship."[2] The professional field of oral history consists of scholars and practitioners from a variety of disciplinary and theoretical backgrounds. What consistently unifies this group is the "material with which it deals": the recorded voice, the interview. We have staunchly defended oral history's relevance and reliability through the decades, yet, with few exceptions, the orality/aurality that defined our material was consistently stripped away in the textual act of archival use and scholarly communication.

Recording technologies enabled the capturing of orality/aurality and, as a result, created and shaped the growth and maturation of oral history methodology and practice. Yet, it has been these recording and playback technologies that have posed many of oral history's greatest challenges over the years. The mechanics of analog audio and video resulted in inefficiencies with regard to access and discovery, constructing a dependence on transcripts, which posed cost-prohibitive and unsustainable models of practice. The constraints of analog delivery systems and distribution and broadcast modalities severely limited public history's applications of oral history. From the archival perspective, oral history proved an exciting and enticing resource to acquire. However, the difficulties posed by time-intensive and financially draining realities of processing oral history collections resulted in an analog crisis in the late 1990s. Hundreds of oral history archives around the United States claimed large collections, but the overwhelming majority of these collections containing thousands of interviews, remained unprocessed, analog, inaccessible, and un-used. In 2006, Michael Frisch began his chapter "Oral History and the Digital Revolution: Toward a Post-Documentary Sensibility" with a sentence that has haunted me ever since:

> Everyone recognizes that the core audio-video dimension of oral history is notoriously underutilized. The nicely cataloged but rarely consulted shelves of audio and video cassettes in even the best media and oral history libraries are closer than most people realize to that shoebox of unviewed home-video camcorder cassettes in so many families—precious documentation that is inaccessible and generally unlistened to and unwatched.[3]

Michael Frisch was correct in simply stating the painfully obvious and largely silent reality pertaining to the actual use of archived oral history. My professional observations, experiences, and conclusions made from working directly with oral history at the collecting, curating, and disseminating phases, was that oral history was relatively easy to collect in large numbers, but it remained an often insurmountable challenge for archives to process in the analog and early-digital context. In the Editors' Introduction to "'I Can Almost See the Lights of Home': A Field Trip to Harlan County, Kentucky," Portelli and Hardy challenge historians to:

> truly explore the full dimension of the sources they create and utilize in scholarship—to engage the "orality" of oral sources. It challenges *all* historians to consider alternative modes of presenting interpretations, modes that render the very act of interpretation more visible while preserving and respecting the integrity of primary sources.[4]

These words were written in 1999, yet analog, textual models are still deeply ingrained and continue to shape the primary modes of oral history expression in the digital age. Despite high praise for their essay in sound, Charles Hardy characterized *Lights of Home* as an "evolutionary dead-end" when asked about its role as a model for digital scholarship. With its fate intertwined with the suspension of the *Journal for Multimedia History*, very few, if any similar works emerged in the following years.[5] *Lights of Home* was not an evolutionary dead-end, but it has taken a significant amount of time for paradigms and infrastructure to catch up. A shift has, indeed, finally occurred, and it is now accelerating. Digital technologies and tools have begun to disrupt the traditional modalities of access, use, and engagement with oral history. This chapter explores, from an archival and a personal perspective, the digital empowerment of the audio and video in the oral history user experience.

When I Hear the Old Men Sing, I Love Ireland More

I came to oral history from a personal background in music and recording, a scholarly background in Folklore, and a professional career as an archivist. I believe in the primacy of the recording in our professional practice, while fully understanding and acknowledging the role of text to facilitate discovery and access. As a young graduate student studying Folklore at Indiana University, I was inspired when I read the writings of Franz Boas, John and Alan Lomax, Leonard Roberts, Sandy Ives, Zora Neale Hurston, Barre Toelkien, and Henry Glassie, model fieldworkers, interviewing, recording, and writing. While reading about the adventures of heroic folklorists, I, invariably, longed to hear the recorded voices that inspired the text.

In 1994, I attended my first course as a doctoral student in the Department of Folklore and Ethnomusicology at Indiana University. The professor who most powerfully captured my attention and imagination that semester was Henry Glassie. Week after week that semester, Glassie, a master lecturer and storyteller, blended engaging stories from all over the world with brilliant slideshows, teaching invaluable lessons and inspiring another group of new scholars. He spoke of vernacular architecture in middle Virginia, Pennsylvania barns, Appalachian musicians, Turkish potters, and Irish historians and storytellers. Fascinated with each new culture and genre I explored in print that semester, I was particularly interested in Glassie's oral history work in Ireland.

Ballymenone is a district located in the north of Ireland in County Fermanagh, and in 1972, it was populated by 42 households. The majority of Ballymenone's residents were Catholic farmers living without the luxury of electricity.

> This was a community that lived by the hearth; most of the roofs were still thatched, while speech, story, and song were the primary forms of social interaction. Folklorist Henry Glassie spent much of his career from 1972 to 1982 in Ballymenone, living among the residents and recording many of their conversations primarily about the past.[6]

Glassie's fieldwork during this time yielded several rich and award-wining ethnographic publications including *All Silver No Brass, Irish Folk History*, and the monumental ethnography, *Passing the Time in Ballymenone*.

During my time in Bloomington, the department received a grant to start the Sound And Video Analysis & Instruction Laboratory (SAVAIL). As part of the exploration process with the newly acquired technologies, ethnomusicologist Dr. Ronald Smith offered a course where a small group of students could utilize the technology to explore innovative and creative ways of analyzing and presenting fieldwork material. As part of my work for the course, I was using an Appalachian folktale called "Wicked John and the Devil" that I had recorded in Wise County, Virginia. Also taking an independent study with Henry that semester, I played him a recording of the folktale. Upon listening, Henry mentioned that Hugh Nolan told a version of the same story called "Coals on the Devil's Hearth." Examining the text of the tale reproduced in Glassie's book *Irish Folk Tales,* I asked the natural question someone producing a multimedia presentation would ask, "Can I hear the recording"? With a contemplative twist of his mustache, Henry glided out of the room and upstairs. Moments later, he appeared with a small packaged reel-to-reel tape in his hand and generously offered it to me for my research. Knowing that this was the only existing copy and that it was potentially fragile, I quickly arranged to digitize the tape. At this point, I had read all of Henry's books from his time in Ireland and so read dozens of stories told by Hugh Nolan. I remember the very moment when I pressed the space bar on my digital audio workstation to hear the newly digitized recording. I distinctly recall Hugh Nolan's voice, his accent, his laugh, his cough, the meow of the cat, the rhythm of the clock, the strike of the wooden match. Suddenly, the words flew from the page and transformed my meaningful engagement with the story. I could almost smell the smoke from Hugh Nolan's pipe.

Henry Glassie had studied at the University of Pennsylvania under folklorist Kenny Goldstein, whose recordings of traditional music, Glassie recalls, "helped to draw me into his profession." Goldstein emphasized recording quality in his training methods, and Glassie took this very seriously: "From Kenny's example I knew how good recordings had to be for public consumption, and with the machinery I had, I did not even try." However, the recording context in Ballymenone was often less than ideal. Glassie continues:

> I did once ask Hugh Nolan to record some stories for me to use as examples in class. The recording was clean, uncluttered, but when I played the tape, my students found it incomprehensible. The dialect was too thick, too alien, the sniffs of his perpetual head cold were too distracting. My tapes, like my photographs, served me as I wrote and no more.[7]

The digitization of this single tape grew into a multi-year effort to digitize Henry's entire collection of audio recordings from Ballymenone, which, over the next few years, evolved into an effort to restore and to publish this field tape collection and to publish select narratives and musical performances from these recordings on CD format.

Over 20 years of typical home storage, the field tapes had begun to audibly degrade. A few were unsalvageable and lost forever. Once I had completed the audio digitization at Indiana University's *Archives of Traditional Music*, the process of selection, editing, and production began. As an early adopter of digital audio recording and production technologies, I did most of the work using *ProTools*, the professional standard at that time for high-end audio production. I remember setting up my digital audio workstation in a room in Henry's house in Bloomington, Indiana, atop a centuries' old wooden table, surrounded by thousands of sound dampening books. This is where much of the final editing and preliminary production work took place. Once the track selections were made, edited, and produced, it was time for restoration and final mastering.

In 1999, I traveled to Boulder, Colorado, to work with master restorationist and mastering engineer David Glasser, at Airshow Mastering. The "restoration" phase of our work was primarily focused on the reduction of "tape-hiss" and the minimizing and removal of some of the other audio artifacts that we had deemed impediments to the listening experience. Analog methods for reducing hiss, such as Dolby Noise Reduction, focused on notching out certain frequencies that corresponded to the noise. Invariably, however, the analog reduction/removal of certain frequencies from the recording would significantly impact the overall recording, often dulling the human voice for the listener. At the time, this was a worthy compromise, as the hiss was far more intrusive. Advances in audio and digital technologies introduced tools to better utilize and deploy *Fast Fourier Transform* (FFT) frequency analysis to more "intelligently" reduce analog hiss. FFT methods and digital audio production tools allow you to sample the noise that you intend to minimize or remove, "analyzes that sample, creates a 'noise-print' and creates an algorithm to mathematically reduce or eliminate the noise—leaving the primary audio signal intact."[8] The result of FFT and its application on Glassie's recordings from Ballymenone were astonishing. Suddenly, the voices and the music became the sonic focal point, and the intrusive analog tape hiss had been successfully minimized.

When David Glasser and I completed our restoration and mastering work, I returned home, proud, excited, and inspired. In 2000, Henry and I traveled to Ireland, back to Ballymenone, to share our production with the few individuals appearing on the CD that were still alive, and with their families. Following breadcrumbs of Henry's field notes, we were able to track down several individual performers and family members, and then we would sit and listen to the recordings. As they listened, I could see the individuals, instantly, transported back in time. Like much of the professional folklore community, I knew these individuals had read Henry's books. But listening, that was clearly different.

In addition to presenting our unreleased CD and connecting with many of Henry's old friends, we recorded more performances. Flute player John Joe Maguire still played with spectacular precision and grace. We met new people along the way in different communities. It was there that I heard the sounds of my first experience beside a traditional Irish hearth. On the plane ride home, listening to my digital field recordings over headphones, I realized, when listening to the sounds of the hearth, that I had made a mistake.

In my article "Noise Reduction and Restoration for Oral History—The Stars of Ballymenone," published in 2012 for *Oral History in the Digital Age*, I wrote about this cautionary tale.

> Not having personally experienced an Irish open hearth full of burning turf ever before, I had no idea that unlike the pops and crackles that one hears from a wood fire, a turf fire hisses. Prior to the CD's completion I found myself in Ireland sitting beside a traditional Irish hearth realizing that the hiss of the hearth resembles, somewhat, the hiss of analog tape. I realized that it was possible that some of the "noise" that we may have worked so hard to minimize, may have overlapped with the natural sounds of the hearth fire and that, at times, we might have been too aggressive with the noise reduction. I identified several instances when this broadband noise that we were taking out was discovered to be, at least in part, the fire—causing serious reevaluation regarding how much background noise we were going to actually take out, and how much we were now going to leave in. Fortunately, we still had time to dial back some of the filters allowing more "noise" back in to the recording, and we restored the hearth back into the recording. This was a very powerful lesson for me in terms of the use of and control of technology in recording and curating oral history.[9]

My experience of returning to Ballymenone with Henry Glassie was profound on so many levels. Even though I had read the books, it was not until I spent hundreds of hours critically and deeply listening to these field recordings, that I felt connected to this community and somewhat responsible for their stories. Production decisions were not just decisions centered on listenability and entertainment. Productions decisions carried deep meanings that needed to be conveyed in the final product.

Indiana University Press was eager to publish this effort; however, they were unsure about just how they were going to package and market it. We had proposed an audio Compact Disc accompanied by extensive liner notes and transcriptions in the CD jacket, but they were uneasy about releasing it as only an audio compact disc. The Press opted for the CD to be packaged with a full sized book written by Henry. Several years later, Henry produced the monumental book *The Stars of Ballymenone*, with the audio CD included.

In a review of the audio CD entitled *Stories and Songs of South Fermanagh* published by the *Oral History Review* in 2007, Gregory Hansen remarked:

> The chance to hear the actual voices of Hugh Nolan, Peter Flanagan, Ellen Cutler and Michael Boyle, wonderfully rounds out the understanding of the words set down in print in Glassie's writing on this remarkable community. Hearing the voices fills in the sense of presence that helps the reader more richly imagine how the storytellers sound and provides a clear and vivid appreciation for the subtle eloquent style of storytelling in Ballymenone... Boyd and Glassie carefully created an audio presentation that accurately expresses how the storytellers, historians, and musicians ideally would wish to have their artistry portrayed.[10]

Henry's research and writings gave voice to the small community of Ballymenone as it underwent violent transformation. However, these voices were previously "heard" only as words on a page. In *Passing the Time in Ballymenone*, Glassie vividly describes a moment at the end of a typical night of drink and music in a nearby "public house."

> With people moving past, laughing, jamming in the doorway, a quiet handsome young man who has known the terror of midnight roads, the engine's hum, the feel of a makeshift bomb in his palms, stands, looking at no one, and says softly, "When I hear the old men sing, I love Ireland more."[11]

The voices and sounds of this community were brought to life by the production and inclusion of the audio CD, the creation and publication of an oral and an aural context for Glassie's written, ethnographic publication. The young man in the pub did not say, "When I read about the old men singing, I love Ireland more," nor did he say, "When I read the lyrics of the songs the old men sing, I love Ireland more." The voices on the audio CD were, indeed, framed by Glassie's scholarly writings, but in many ways, the voices profoundly transformed Glassie's scholarship.

If a Tree Falls in the Archive, and There Is No One There to Hear It, Does It Make a Sound?

I stepped into professional oral history in 1998 when I was hired as the senior archivist for the oral history and folklife collections at the Kentucky Historical Society, home of the Kentucky Oral History Commission. While attending my first Oral History Association annual meeting in 1998 in Buffalo, New York, I first learned of Project Jukebox, heard Charles Hardy and Alessandro Portelli present the rough cuts of "I Can Almost See the Lights of Home" in a darkened conference room, and later listened to professional icons such as Sherna Gluck, Linda Shopes, and Terry Birdwhistell discuss and debate putting oral history interviews up on the Internet.

Returning to Kentucky and to my desk that following Monday morning, I was, all at once, inspired and frustrated. I knew that I had core competencies in digital recording and digitization from earlier experiences as a musician. I was fascinated by emerging technologies and could only dream of potential applications to oral history. Yet, I recall gazing over the thousands of catalog records at my fingertips and knew that only a small percentage of these cassettes had ever left the shelf. While oral history had such potential to "breathe life into history" the realities and limitations imposed by analog technologies and scholarly conventions left that potential unfulfilled. I returned to my desk that week having the same discussion in my head about the role of oral history's audio and the preference for transcripts debated decades prior. What bothered me most was when Louis Starr bluntly stated in 1977 that researchers much preferred to interact with oral history transcripts by ratios of "a thousand to one and higher," than to listen to the audio or watch the video—and he was correct.[12] He was still correct in the 1980s when Dale Treleven added that researchers still preferred to "'listen' to a typescript."[13] Starr was still correct in 1998 when I began working for the Kentucky Oral History Commission, and, in some ways, he is still correct today. Even in today's digital context, users and researchers of oral histories in archival settings largely prefer the transcript as an access point. In the typical reference encounter I still hear on a regular basis "just send me the transcript," or a surprised, "Oh, it's not transcribed?" In the analog context, the typescript transcript not only made search and browse much more efficient for the researcher, it removed the need for, and dependence on, a machine to achieve successful human engagement with archived oral history. Computers have brought the machine dependence back, but the difference today is computers are now part of our everyday lives. As a researcher, I too love having a transcript as a discovery tool, to quickly search and seek information. When researching, rarely do I read a transcript in its entirety.

As an oral historian, I would like to assume that people working with oral history in an archival setting would be drawn, first, to the recorded audio and video. As an archivist, however, experience has shown that it is idealistic and naïve to think users can effectively discover specific information in interviews without the assistance of a transcript or an index.

The audio player (both analog and digital) alone poses a cumbersome, inefficient course for a user's experience in discovery and interaction with information. Text provides cues and visual navigation and creates access points to audio and video. We cannot effectively browse audio and video resources, for no sooner do you speed up or skip a passage, you have, potentially, missed vital information. Over the years, archives have struggled to find analog and digital discovery and navigation solutions for oral histories in the absence of a transcript. The tape-log was a popular "compromise" which typically included a combination of elements including a partial transcription, keywords, and narrative description of content, marking the location of the content to the correlating counter number on the cassette or reel-to-reel machine. The tape-log could be created more efficiently and affordably than the verbatim transcript, and it served as a finding aid for researchers, a way to browse the content of an interview and a guide to the location of the content residing on magnetic media.

In his article "Oral History, Audio Technology, and the TAPE System," published in 1981, Dale Treleven describes an innovative effort at the Wisconsin State Historical Society to "combine the capabilities of the new audio technology with vastly improved recording tape to develop more efficient procedures for the processing, preservation, and retrieval of information contained on taped one-on-one interviews."[14] The TAPE or "Timed Access to Pertinent Excerpts" system harnessed the potential posed in "two-track stereophonic" cassettes to assist researchers in navigating the audio. Treleven outlined a system that required:

1. The creation of an audible "pre-recorded time-signal" occurring at five second intervals on one channel of the master tape.
2. The creation of a user copy.
3. Time-coded abstract is created while listening to the user copy. Abstracts included segment time, titles, and descriptions.
4. Preparation of index cards for "names, proper nouns, historical phenomena, concepts, etc., which appear in the abstract" and the preparation of a final index.
5. Drafting of an "Interviewer's Introduction" to help frame the context of the interview.
6. Catalog Cards.[15]

Today, the TAPE system seems cumbersome and awkward. In its time, this was a brilliant attempt to make discovery, access, and use of oral histories more efficient and cost-effective. TAPE was built on the assumption that "orality" was central and that the users of oral history in the archival context would benefit greatly from engaging with the audio recording itself. Treleven firmly asserted:

> Neither a typescript nor any other written finding aid is a substitute for the sound recording itself. In the case of an oral history interview, the tape contains the most accurate simulation of the words actually spoken when an interviewer and interviewee allow their questions and observations to be captured by a mechanical recording device.[16]

Treleven's efforts at Wisconsin represent a pre-digital, pre-internet innovation intentionally asserting the recording in a framework to assist the researcher. While this system was successful as a short-term solution, it was not sustainable. It still suffered from the analog constraints requiring physical presence and laborious, manual efforts to navigate content. Using oral history collections in the archive still required significant and often unrealistic commitments of time and effort.

Digitization and the Internet offered great promise for releasing the traditional analog access constraints on oral history and for raising the expectations of researchers. From a discovery standpoint, collection and interview-level metadata could easily be made searchable online. One of my early challenges at the Kentucky Historical Society was to convert the printed *Statewide Guide to Kentucky Oral History*, a collection-level guide to oral history collections housed at over 40 different repositories throughout Kentucky, into an online searchable database. In 2001, we proudly launched the online guide. The online guide had a robust search and browse interface but it had one major limitation: it provided the users an incomplete pathway to what they were hoping to access, the interview itself. To hopeful users, it was like a sidewalk that just ended. Users, invariably, would call and ask, "How can I access the interview itself?" When I repeatedly explained that the tool was not a digital object repository, that it was an online finding-aid to collections throughout the state, and that they would still have to contact the individual archive in order to inquire as to how to access the interview, the users invariably hung up, disappointed. The facilitation of online discovery raised users' digital expectations of access, only to swiftly quash those expectations with harsh, analog reality.

Transcribed interviews were relatively simple objects to present in early online environments, however researchers and users had to, still, physically interact with the audio recording. Early Internet delivery systems rarely paired the audio recording with the transcript. In part, this was a bandwidth issue. Truly streaming media was not readily accessible and affordable for archival institutions, and oral history interviews tend to be lengthy audio recordings. Lengthy recordings made for large file sizes, which led to unusable download speeds. Digitization of audio and video collections greatly lagged (and continues to do so) in contrast to paper-based and photograph collections. The complexities of digital or digitized audio and video in the late 1990s presented a major barrier to online access to the recorded sound and video. Louis Starr's characterization of the transcript, from the user's perspective, remained true as we approached the end of the century.

William Schneider's *Project Jukebox* and Sherna Gluck's *Virtual Oral/Aural History Archive* were creative and innovative digital models for creating early solutions to oral history's analog access challenges, and both were conscious attempts to reinsert orality into the oral history archival process. Additionally, Schneider and Gluck were able to utilize linkable data to construct contextualized frameworks to create highly curated user experiences, and as a result, more effectively connect to the individuals and communities being represented in their oral history projects.

Both Project Jukebox and VOAHA were breakthrough moments for the digital era of oral history. Both web interfaces utilized hypertext and direct linking to structure the user experience. Both included search and browse experiences and, most significantly, both user experiences culminated in the digital presentation of audio, in the form of organized excerpts. In his interview for this book, Schneider recalls earlier phases of Project Jukebox and the typical user experience:

> One of the things we were very concerned about was people being able to navigate in multiple ways throughout the program, enter the program in multiple ways. So in the case of something like the *Yukon Charlie* one, the most obvious entrance, in a way, is by searching under a person's name. If you're from the community, you might want to search under a person's name. But if you're not from the community, you might just want to search under the name of the park, and you could do that. But you also might want to search under a theme, such as D-2, which refers to a section of the Alaska Native Claims settlement act and the emergence of these park areas. Or you might search under administrative history or something else that would lead you to an entry into that program. In the case of Yukon Charlie, the added material, with photographs and stuff from Brad Snow, later on in the program, it gives the viewer a picture of the life of the river people. So if you're studying back to the land folks, people that had decided that they wanted to go into the wilderness and find, and make their way in a subsistence way, then you could do those types of searches once you got into the program itself. So there were multiple points of entry, and navigation, once you entered this site, and that's the beauty of that architecture.[17]

In his chapter for this book, Schneider candidly admits to funding issues and the resulting need to automate contextualized access as being the initial motivator for the development of *Project Jukebox*. Recognizing that a contextualized approach to framing oral histories in an archival setting was an incredibly laborious and expensive undertaking and that archival budgets seemed to be declining rather than expanding, Schneider looked to automated access as the sustainable archival workflow. The result was, indeed, a success in automating contextualized access to oral history collections. The legacy of *Project Jukebox* was a dramatic innovative digital model for the delivery and presentation of oral narratives, in the context of a community.

For Sherna Gluck, her motivation was, from the beginning, to provide unmediated access to feminist voices. For the user entering through the front gates of VOAHA, each navigational step—the front page, topical browse, collection, and interviewee lists—down to the individual excerpt, provided description, context, and choice. If the user were so inclined, they would click "play this segment" and be presented with RealPlayer and the corresponding audio excerpt. In her oral history interview, Gluck describes her early motivation for VOAHA:

> Well I, as a lot of people know, had argued vociferously against transcripts when I first got involved in 1972. And it was not just a resource issue. It was, we had a community based feminist history research project. But I really wanted the way in which we were using oral history and what was becoming a feminist oral history movement was trying to bring those voices to the community...I was really from the start focused on orality.[18]

Later on, Gluck specifically reflected on efforts to contextualize the individual segments and to feature the "significant" content.

> I've been concerned and remain concerned about how people use oral histories for purposes for which it was not intended...we were trying to figure out a way to make it useful and to focus on things that we felt were significant in any segment...I just felt after going for years, from '72 to 2000, having people using my own interviews, sometimes, initially, they'd come up to my house because that's where they were. Trying to figure out how to make them usable.

The development of the user interface for both projects were tremendous undertakings, requiring careful curation and laborious description of individual excerpts, built on a complex information architecture of links and dependencies, and framed by an attractive user interface. Both emerged as new models for the archival user experience. Users of oral history collections were, for the first time on this scale, empowered to efficiently search, explore, and engage with large collections of oral history in an online research context.

In subsequent years, a few similar online oral history resources began to emerge. My initial experiences with Project Jukebox and VOAHA dramatically shifted my perception of the oral history archival reality. I quickly came to the conclusion that the Internet was useful for so much more than searching for an online catalog or the static delivery of transcripts. Additionally, I grew frustrated by the user experience I was providing with our reference model. For the oral history collection at the Kentucky Historical Society, a rich collection of over 6,000 interviews at that time, we provided only vague collection-level metadata, few transcripts and an extremely limited access experience for analog audio. This frustration intensified as our award-winning *Civil Rights Movement in Kentucky Oral History Project* progressed, culminating in the production of *Living the Story*, a documentary derived from the oral history project. The publicity and attention around the documentary and the project greatly increased demand for access to the archival collection, yet the only element of the project I could easily provide at the time was the transcript. Driven by a motivation to enhance access to our collections in such a way that inserted the voices into the research engagement with oral history, I began work on the *Civil Rights Movement in Kentucky Oral History Project Digital Media Database*, an online digital project which offered themed and geographical browsing of content, global search, and a user experience that drilled down to organized audio excerpts as well as linking to online, searchable transcripts. I would have linked to the entire audio interview if the Kentucky Historical Society had access to streaming technologies at the time. Since they did not have affordable access to a streaming server, I gravitated to the *Project Jukebox* and *VOAHA* models of the database driven, online audio excerpt.

In retrospect, the online *Civil Rights Digital Media Database*, as well as *Project Jukebox* and *VOAHA* were not archival repositories; they were complex, elaborate, and beautiful digital exhibits offering a curated, even guided, user experience, connecting online users to powerful audio-visual content. Even while in the style development phase of design for the project, I knew the success of the *Civil Rights Movement in Kentucky Oral History Project Digital Media Database* was in the oral history user experience that it provided. The site was gorgeous, the information architecture was logical, very much designed in response to archival demand, and it connected users, within moments of entering the site, to the voices. I measured success by use. In a pre-Google Analytics environment, use was measured by feedback, and the feedback we received was overwhelmingly positive. Interviews from this collection were being utilized in numerous secondary and postsecondary classrooms around the state. Like the users of *Project Jukebox* and *VOAHA*, our users were not just reading stories about living under segregation in the South, they were actively engaging with audio and video. The beauty in the model Sherna Gluck and William Schneider created for online oral history-driven, boutique digital projects was in the framework of a user experience that provided engaging and efficient access to contextualized selections in the oral history interviews. From a digital projects perspective, these works were brilliant models that are emulated, even today.

From the archival perspective, however, this boutique approach—the manual creation of audio segments, the manual construction of links, the detailed description, contextualization and tagging of the audio segment—has proven, time and time again, unsustainable for large-scale oral history archival workflow. In addition to constructing workflows that were difficult for others to emulate on a large scale, each project suffered greatly from rapid obsolescence. Digital platforms are constantly updated, upgraded, and eventually replaced. In addition to the oral history interviews, technical infrastructure and design must also be maintained and curated on an ongoing basis. As the passage of time progressed, all three institutions behind these innovative, model digital projects sought grants to resurrect the online resource. In my chapter "Achieving the Promise of Oral

History in a Digital Age" in the *Oxford Handbook of Oral History*, I reflect on my experience:

> A more scalable model was needed for delivering the majority of our digital assets in a less labor intensive and less expensive fashion... A database that was originally designed for dynamism and rapid updates ultimately became a static exhibition.[19]

Following my departure from the Kentucky Historical Society in 2006, the *Civil Rights Movement in Kentucky Oral History Project Digital Media Database* was digitally abandoned, opened up to online hackers and eventually taken down. After some time, the online resource returned to life; it has yet to be updated and at this point, it remains incompatible with many modern Internet browsers. *Project Jukebox*, *VOAHA*, and the *Civil Rights Movement in Kentucky Oral History Project Digital Media Database* served as early and inspirational models for creating powerful frameworks for an online archival interface connecting researchers to the "voices" in large oral history collections, but they also serve as powerful cautionary tales, imparting lessons learned regarding archival workflows, obsolescence, and sustainability.

As an oral historian, folklorist, and archivist who has primarily worked with archived oral histories for my entire career, I have become firmly committed to the ideal that the oral history community cannot structure our fundamental access workflows and strategies on models that require unrealistic amounts of continually escalating funding. In today's innovative digital climate, it seems that you can do just about anything with a grant. What you cannot do, necessarily, is sustain what you created with that grant, after the grant funding runs out. The accessibility and affordability of digital audio and video recording technologies, combined with an increasing popularity in oral history methodologies have greatly intensified the need for affordable and sustainable strategies in providing an enhanced user experience, which, in my mind, necessitates connecting users to the recorded interview.

Several interesting and innovative resources emerged, offering exciting presentations of oral history in a digital environment. These isolated examples typically enhanced access to oral history in incredible ways for a relatively small, targeted group of interviews, but were often built on proprietary technologies. Most were grant funded, un-replicable without comparable grant funding, and proved unsustainable following the end of the grant period. Despite the beginnings of the digital revolution in oral history, the same problems were present: the majority of oral history interviews housed in a typical archival institution, funded by a reasonable budget, utilizing contemporary archival workflows, and best practices were still difficult to discover and even more difficult and laborious to use in the absence of the transcript. In the larger library and archival context of massive digitization efforts, major online repositories of digitized and searchable photographs, manuscripts, books and newspapers, oral history—especially the audio and video interviews themselves—remains noticeably absent. In a pedagogical context that emphasizes the use of primary sources in classrooms ranging from elementary school to higher education, with few anecdotal examples, archived oral histories remain underutilized. Oral history interviews continue to be a difficult information package to accession, process, discover, and use. The voices of oral history, the "material" of our professional practice, for the most part, remain silent.

In 2008, I became the director of the Louie B. Nunn Center for Oral History at the University of Kentucky Libraries. Looking at a largely analog collection containing thousands of audio and video interviews, an emphasis on collection level metadata, minimal online presence, and relatively very few transcripts, I intensified my personal commitment to providing sustainable models to connect archival users to the online primary sources. I watched the Nunn Center users and researchers gravitate more to collections that were transcribed. Despite our standard warnings to corroborate direct quotations with the original audio or video interviews, I watched researchers quote and misquote from transcripts that were, often, not even verbatim representations of the text. In general, our audio and video interviews remained on the physical and virtual shelves. I do not believe that researchers generally wanted to ignore the audio and video interviews because they were lazy and uninspired by the human voices telling the stories. Time-based media in both the analog and digital realm is difficult and time consuming to use. Contemporary digital archive and library frameworks have been constructed to optimize the user experience for repositories of digitized text and images, and they have generally failed in providing usable architecture for enhancing the users' experiences with online audio and video.

Wary of outmoded and ineffective archival models for providing access to oral history collections, in 2009, I designed and led the team to construct and launch the first version of the Oral History Metadata Synchronizer (OHMS) to enhance access to online oral history. Drawing from lessons learned in the analog and digital past, OHMS was created as a web-based digital tool to integrate the audio and video interview recordings in their entirety, with the searchable and readable text that users crave and demand. Common digital platforms, such as CONTENTdm, allowed for simultaneous presentation of an oral history's recording and text; however, the two elements remained unconnected. The user could search a transcript but they still had to manually locate and navigate to the corresponding moment in the audio or video recording. Seeking a way to break through this lingering analog constraint on access, the initial version of OHMS simply connected a textual search of an online transcript with the corresponding moment in the online audio or video. I was feeling pretty proud of my accomplishment and made the rounds at conferences discussing our innovative approach. However, two major problems soon emerged. The first problem was that OHMS required transcripts. If the Nunn Center transcribed every interview that we accessioned last year alone, we would have spent over $250,000 on transcription. As oral history methodology becomes increasingly popular, I can no longer keep pace with the funding necessary to transcribe on a mass scale. The second, more profound problem was that OHMS was designed and built to work only for our in-house system. It was the solution for the *Nunn Center's* oral history collection but, as originally designed, was of no use to outside institutions.

In early 2011, disillusioned by our escalating and unrealistic cost of transcription, I transformed the OHMS system to include indexing of audio and video. The OHMS Indexing Module enables the efficient annotation of an interview, including the creation of a title, partial transcript, segment synopsis, keywords, subjects, GPS coordinates, and related external hyperlinks for each "segment" of the index. As it did with the online transcript, the OHMS Viewer connects the online user to the correlating moment in the audio and video. In fact, with the incorporation of GPS coordinates and hyperlinks as metadata elements, the user can connect to locations on Google Maps or link out to related photographs or websites to contextualize the passage that continues in the background. I have come to embrace and advocate the *Frischian* notion that, in many ways, the index is advantageous for mapping the natural language of speech to meaningful concepts. But as a researcher, I also appreciate the granular search provided by the transcript. As an administrator, I prefer the budgetary opportunities created by indexing. However, users and researchers engage with oral history in different ways. I have come to the conclusion that the better user experience is when the OHMS viewer presents both an OHMS Index and a transcript, allowing the user to toggle between both discovery tools as needed. Nevertheless, the archive no longer has to hold interviews hostage on the physical or virtual shelves for decades while awaiting funding for transcription.

In many ways, the OHMS Indexing Module was modeled on the analog tape logs that I created and curated in my early years as an archivist, as well as on the TAPE system developed by Dale Treleven. As was the case with the Wisconsin Historical Society, OHMS indexing has greatly increased access to our oral history collections—access to the audio and video—for a fraction of the cost of transcription. However, the most significant result of the creation and implementation of OHMS is not the cost savings afforded by indexing. OHMS places the audio and video out in front of the text, allows a contextualized navigation of an online interview, and includes a transcript (if available) as well as an index that effectively maps natural language to useful concepts. Oral history contains a massive amount of information that can prove cumbersome to navigate, yet to the researcher, it is the individual moments in an oral history interview that matter. OHMS empowers online users of oral history to search, explore, and connect to those moments on a large scale, effectively and affordably.

Additionally, OHMS has become an exciting outreach and engagement tool for the Nunn Center. After a year of using OHMS as an indexing tool, it became clear to me that indexing or annotation of the audio and video interviews was an engaging experience. I can walk in to most rooms and in a few moments generate excitement over conducting an oral history project. I can, and do, train volunteers to run equipment and to conduct professional quality interviews. What I could never get volunteers for is the accurate, professional-level transcription. The OHMS Application is web based. With a wireless network and a reasonable Internet connection, you can index oral history interviews from the back porch on a nice afternoon. Additionally, I realized that an individual did not require a graduate degree in Library and Information Science in order to create a meaningful index.

As an experiment, using untrained "indexers," I identified one of the Nunn Center's greatly underutilized collections focusing on the desegregation of Major League Baseball. The Nunn Center had 75 interviews with numerous players, coaches, and journalists who witnessed and participated on some level of this historic transformation. The collection was not transcribed and, therefore, rarely was used. I was teaching a graduate course on oral history and decided that these students, who admittedly knew very little about the history of baseball, would index this collection. OHMS does allow you to upload any thesaurus or controlled vocabulary and suggest those terms as keywords or subjects as the indexer types. We created an extensive thesaurus containing personal names, stadiums, official team names, as well as esoteric baseball terms. Students were instructed to prioritize terms in the thesaurus, but in the absence of a suggestion from the controlled vocabulary, to be descriptive and comprehensive. Within a three-week period, all 75 interviews had been indexed. The indexes were not perfect, but they required minimal cleanup by Nunn Center staff before launching the collection online. Shortly after, the entire collection was made available using the OHMS Viewer, for free. Since that initial experiment, we have collaborated with numerous professors and students, as well as with communities and organizations, to begin indexing interviews. Although we are early in this process, there is great promise in engaging communities to assist in our efforts to make oral history more accessible.

OHMS Indexing solved the first problem I posed. The second problem was that OHMS was initially built on proprietary platforms and constructed to only work on our system. Later in 2011, the Nunn Center received a National Leadership Grant from the Institution of Museum and Library Services (IMLS) to make OHMS an open source and free solution. OHMS is designed to minimize dependencies and be universally compatible with online content management systems such as Omeka or CONTENTdm. OHMS was constructed to utilize ubiquitous formats such as XML and CSV, both of which can be opened and manipulated using a basic spreadsheet or word processing software package.

Furthermore, OHMS creates an information package that, of course, works effectively with the OHMS Viewer, but will easily and seamlessly map to and integrate with future systems as well. What results from OHMS is a sustainable archival solution for providing enhanced online access to oral histories, and the ability to effectively and efficiently navigate, discover, and engage the orality and the content of our oral history materials in a flexible and affordable online environment.

As the Nunn Center has evolved these past five years, our oral history interviews—the audio and video recordings—are being used in classrooms and by communities on an unprecedented scale. The Nunn Center used to report hundreds of interviews being used each year. This month alone, we have logged over 8,000 online engagements with our oral history collection from all over the world. As a field, however, oral history still has a long way to go in providing large-scale, effective access to our interviews and collections. The archival community must work together with the oral history community to explore, adapt, and innovate in order to better serve our mission of preservation of and access to oral histories. Inaccessibility in an Internet age means obscurity and irrelevance. We need to embrace what Rob Perks refers to as:

> a new model of partnership working between oral historians—both in the community and in higher education—and archivists, where both can create and document contemporary culture in a collaborative relationship.[20]

Decades-old struggles to make oral history more usable and accessible have succeeded in the short term, yet mostly failed to change analog paradigms of access and use. Free technologies and platforms such as YouTube, Vimeo, SoundCloud, Omeka, Drupal, Wordpress, and OHMS are now at our fingertips. We must quicken our transition, our mindsets and paradigms, and our archival workflows and procedures to adapt and accommodate users' expectations. When we do, our interviews will be used.

The role of the archive in the digital age is changing and its importance rapidly increasing. Interviews, designed and readied for access, must be made more accessible and usable. Interviews that need temporary restrictions must be better protected. Archival access to oral history needs to be granular and precise in order to be most useful, yet, admittedly, budgets continue to shrink. Like William Schneider and Sherna Gluck, we need to explore and implement scalable solutions to better merge and integrate the metadata (description, index, and transcript) with the audio and video recordings. The dreams of *Automatic Speech Recognition* (ASR) and its future applications for oral history are big, as are the present limitations of the technology. Archivists and oral historians need to strive to enhance access to their oral history interviews and collections using tools such as OHMS to create effective, efficient, and sustainable models for automated access. We need to encourage open access and usage of the recorded interviews where and when appropriate and make oral histories accessible using platforms such as the *Digital Public Library of America* (DPLA). Scholarly publications should not just be quoting and citing oral history interviews, they should be regularly embedding multimedia and linking out to both the interview and to the specifics moments—placing the primary sources at the center of scholarly discourse. Ahead of their time, Charles Hardy and Alessandro Portelli demonstrated this so meaningfully with *Lights of Home* in 1999.

The field of History, currently, plays a very minor role in current scholarly discourse regarding the "Digital Humanities." As I have personally observed through conversations, scholarship and ongoing hiring trends in professional circles pertaining to the maturing, yet allusive digital humanities, the voices of History remain somewhat quiet. In a 2014 blog post, Stephen Robertson stated, quite frankly, that the term "dh" is "clearly standing in for digital literary studies."[21] With some important exceptions along the way, History's promise and impact on the digital dialogue remains somewhat unfulfilled. Yet Oral History has the distinct advantage of the human voice and the still and moving images, and the power to engage and move listeners in profound ways. The voices and moving images that make up archived oral history have always had the power to "breathe life into history," to re-quote Paul Thompson, but we must breathe new life into our workflows and infrastructure.

About a year ago, my oldest daughter was helping me clean out my office at home. Opening up a shoebox of old audiocassettes, I was stunned when she held one up and asked, "What is this?" I proceeded to try to explain analog access to a child who is conditioned by an instant access, on demand user experience delivered by the iPod. I explained how we had to forward, rewind, and wait—then stop, play and then forward or rewind a little more—and then wait once more. She shrugged her shoulders and poignantly said, "I just want to click on it to listen." Her words resonated with me in ways she will never fully understand. Her brilliant and concise observation was, in my mind, a partial fulfillment of Louis Starr's 1977 conjecture that "future generations may prove more aurally oriented."[22] The archival community must strategically adapt workflows and capabilities to create solutions that accommodate the users of our oral history interviews, because if we build platforms that make it easy to do, they will indeed, choose to "click to listen."

Notes

1. Paul Thompson, *The Voice of the Past* (Oxford: Oxford University Press, 1978), 18.
2. Kenneth S. Goldstein, *A Guide for Field Workers in Folklore* (Hatboro, PA: Folklore Associates, 1964), 1.

3. Michael Frisch, "Oral History and the Digital Revolution: Toward a Post-documentary Sensibility," in *The Oral History Reader*, 2nd edition, ed. Robert B. Perks and Alistair Thomson (London: Routledge, 2006), 102.
4. Charles Hardy III and Alessandro Portelli, "Script," " 'I Can Almost See the Lights of Home': A Field Trip to Harlan County, Kentucky," *The Journal for Multimedia History*, 2 (1999), http://www.albany.edu/jmmh/vol2no1/lightsscript.html.
5. Charles Hardy III, Interviewed by Doug Boyd, April 30, 2014, available online at http://www.digitaloralhistory.net.
6. Doug Boyd, "Case Study: Noise Reduction and Restoration for Oral History–The Stars of Ballymenone," in *Oral History in the Digital Age*, ed. Doug Boyd, Steve Cohen, Brad Rakerd, and Dean Rehberger (Washington, DC: Institute of Museum and Library Services, 2012). http://ohda.matrix.msu.edu/2012/06/noise-reduction-and-restoration-for-oral-history/.
7. Henry Glassie, *The Stars of Ballymenone* (Bloomington, IN: Indiana University Press, 2006), 502–503.
8. Boyd, "Case Study: Noise Reduction and Restoration for Oral History–The Stars of Ballymenone." http://ohda.matrix.msu.edu/2012/06/noise-reduction-and-restoration-for-oral-history/.
9. Ibid.
10. Gregory Hansen, *Review of The Stars of Ballymenone*, Henry Glassie. *Oral History Review*, 34(2) (Summer–Autumn, 2007): 147–149.
11. Henry Glassie, *Passing the Time in Ballymenone: Culture and History of an Ulster Community* (Bloomington, IN: Indiana University Press, 1995), 85.
12. Louis Starr, "Oral History," in *Encyclopedia of Library and Information Sciences*, Vol. 20, ed. Allen Kent et al. (New York: Marcel Dekker, 1977), 443.
13. Dale E. Treleven, "Oral History, Audio Technology and the TAPE System," *The International Journal of Oral History*, 2(1) (1981): 26.
14. Ibid., 32.
15. Ibid., 33.
16. Ibid., 34.
17. William Schneider Interview.
18. Sherna Gluck Interview.
19. Doug Boyd, "Achieving the Promise of Oral History in a Digital Age," in *The Oxford Handbook of Oral History*, ed. Donald Ritchie (Oxford: Oxford University Press, 2011), 294–295.
20. Robert B. Perks, "Messiah With the Microphone? Oral Historians, Technology, and Sound Archives," in *The Oxford Handbook of Oral History*, ed. Donald Ritchie (Oxford: Oxford University Press, 2011), 327.
21. Stephen Robertson, "The Differences Between Digital History and Digital Humanities" *Dr Stephen Robertson* (Blog), May 23, 2014, http://drstephenrobertson.com/2014/05/23/the-differences-between-digital-history-and-digital-humanities/.
22. Louis Starr, "Oral History," in *Encyclopedia of Library and Information Sciences* Vol. 20, ed. Allen Kent et al. (New York: Marcel Dekker, 1977), 443.

PART II

Discovery and Discourse

CHAPTER 5

Beyond the Transcript: Oral History as Pedagogy

Marjorie L. McLellan

From Folklore to Oral History

When I recorded my first interview, I was a graduate student in American Folk Cultures in the Cooperstown Graduate Program. I interviewed Lithuanian Americans in Amsterdam, New York, about traditions surrounding food and celebration. The following year, in February, at the northern tip of Wisconsin's Bayfield peninsula, I interviewed Finnish Americans about everyday life for Old World Wisconsin, the then-new outdoor museum that brought together buildings representing the many ethnic groups living in rural Wisconsin. I drove along roads buried deep in snow to snug log houses with wood-burning stoves where I was greeted with thick black coffee, rye bread, pickled sardines, fruit soup, other local foods, and stories.

I was an applied folklorist, with a cassette recorder and a camera, charged to document everyday life for museum staff that would heat a sauna, plant a garden, furnish a house, and make yogurt. By the following summer, museum interpreters in period clothing were doing many of the tasks that I had documented. It would be more than a decade and well over 100 interviews later before I considered myself an oral historian.

At Old World Wisconsin, a collection of glossy 8" × 10" prints from a cache of glass plate negatives helped the staff furnish the museum's two *fachwerk* or half-timber Pomeranian immigrant houses and lay out the gardens, fruit trees, outbuildings, and fields. The prints were made from 4" × 5" glass plate negatives supplied by Edgar Krueger who grew up and, in his seventies, still lived on a nearby farm. The photographer—Edgar's father, Alexander Krueger—seemed to have taken up photography at about the time he, his wife, and their young twins, Edgar and Jennie, shared the farm with Alexander's parents, sister, grandfather, and step-grandmother in the late 1890s.

Curious about the photographs of family members in very traditional clothing and wooden clogs standing with spinning wheels and hand-woven baskets in front of a brick and timber barn, I made an appointment to interview Edgar Krueger. Working at Old World Wisconsin, I was interested in everything—how they lived and worked, why they had dressed in such conservative clothing (they were dressing up), and what other photographs he could share. Krueger showed me over 800 glass plate negatives, each neatly jacketed and identified, along with family albums of postcards—more family photographs printed and sent back and forth between siblings and cousins.

Armed with Krueger family stories, a three-ring binder of photographs, and some historical research, I set off for Washington, DC, where, helped generously by Alan Jabbour and Carl Fleischhauer at the Library of Congress' American Folklife Center, I met with staff at the National Endowment for the Arts (NEA) and the National Endowment for the Humanities (NEH).[1] With encouragement from NEH, George A. Talbot, director of the Wisconsin Historical Society's Sound and Visual Archives—known as Iconography or Icon—agreed to work with me on a proposal for a traveling exhibit about family and memory. I was in awe of Talbot's rich and insightful exhibit for the US Bicentennial, "At Home: Domestic Life in the Post-Centennial Era, 1876–1920," then on display in Madison.[2] We spent many weekends sitting in his garden, drinking coffee on the University of Wisconsin terrace by Lake Mendota, or wandering around garden supply stores, drafting and redrafting a planning grant, only to see it rejected. We revised and successfully resubmitted the proposal to NEH.

Over time, I interviewed the Kruegers and their kin across the Midwest and collected four microfilm reels of documents and images and thousands of negatives along the way. Dale E. Treleven taught me to record interviews at the Wisconsin Historical Society. He saw the audio recording that captured nuances of emotion, dialect, and style as the primary source. He had developed a system to index analog recordings: the Timed Access to Pertinent Excerpts (TAPE) system involved duplicating the recorded interview on one channel of a stereo recording with an audio time code on the second channel.[3] The time code overcame the inconsistency of locating excerpts by the counters that tracked the number of revolutions on tape recorders. We produced tape summaries for the Krueger interviews, providing a topical heading and summary for approximately each five-minute segment of the recordings so that a researcher could access and listen to the exact point in the recoding. Transcripts were generally produced only to aid analysis or for publications and exhibits.

In the exhibit itself, Talbot and I sought to preserve the orality and performance through the use of video and audio recordings. Concerned about technical quality over the life of the exhibit, we produced one of the first generation of optical videodisc or LaserDisc recordings made at 3M.[4] Our exhibit, *Six Generations Here... A Farm Family Remembers*, opened at Old World Wisconsin in 1982.[5]

Our humanities advisors—Kathleen Conzen, Robert Ostergren, Daniel T. Rodgers, Kenneth Ames, and Carl Fleischhauer—raised fascinating questions and perspectives about the research data.[6] As a result of working with them, I went on to earn a PhD in American Studies at the University of Minnesota. Over those years, I co-chaired an American Folklore Society annual meeting, produced public radio shows about Wisconsin folklore, worked on television documentaries about families engaged in fishing and farming, developed a school-based project around the diverse cultures and traditions of Minneapolis families, and taught "A Place to Call Home," a teacher institute on Minnesota families grounded in an edited selection of primary documents from family collections at the Minnesota Historical Society.

On the way to my doctorate, I had studied cultural geography, religion, folklore, history, and cultural anthropology. My dissertation co-advisors were anthropologist Riv-Ellen Prell-Foldes and historian Elaine Tyler May, and my approach was influenced by works by, and/or classes with, geographer Yi-Fu Tuan, folklorist Gary Alan Fine, historians Allen Isaacman, John Modell, and Sara Evans, and visiting scholars including Juha Pentikäinen, Victor Turner, and Barbara Myerhoff.[7] The resulting dissertation, *Bread and Glue: Tradition in the Lives of Four American Families,* is an ethnography of families and memory.[8]

The threads that connected these disparate experiences and contexts were my interests in public humanities, personal narratives, and the ways that families imagine and document themselves and their shared histories. I conducted ethnographic field research—time spent interacting with people and observing stories told in context—as well as archival research. Approaching recordings as a folklorist, I listened for the personal and family narratives embedded in the interviews. Narratives emerged from the stream of responses about soldiering in the South Pacific, driving a bookmobile, giving birth before the midwife arrived, migrating from Oklahoma, learning to play a banjo, or organizing for the union. A recorded interview captures stories and performance, albeit out of context like a fish out of water, and in my research, the interview rarely stood apart from other forms of documentation.[9]

Oral History as Pedagogy

By now an interdisciplinary scholar and applied folklorist with experience teaching American studies, composition, and urban studies, I was offered a tenure-track position in the Department of History at Miami University, teaching on the Middletown regional campus. The department chair, Allan Winkler, urged me to build local ties, and that became my mission. I attended the Organization of American Historians annual meeting where the jovial Charles Morrissey, a remarkable interviewer and teacher, encouraged me to join the Oral History Association (OHA).[10] Like the American Folklore Society, OHA intentionally wove together the threads of academic and public or applied work. Scholars, teachers, and practitioners shared a deep appreciation for research with and about ordinary people and everyday lives. I attended the 1992 OHA annual meeting and joined the Education Committee; the following year I became the committee chair.

Influenced by Watts and Freeman's *Generations: Your Family in American History* (1988), I had already integrated the discussion of family narratives and documents—diaries, memoirs, letters, and oral histories—into history courses. By recording and transcribing oral history interviews, students participated in the work of historians. An interview is far more than a simulation—it is a step in the process of learning the craft of historical research. I created an assignment that empowered students to become co-authors of archival historical documents and to examine personal agency in history, analysis and use of historical evidence, continuity and change in everyday life, and the impact of historical developments on ordinary people. At the same time, the assignment addressed the growing expectation to promote "writing across the curriculum." However, oral history offered something more. Published in 1990, Ernest L. Boyer's report, *Scholarship Reconsidered: Priorities of the Professoriate*, exhorted faculty to value the scholarship of integration or synthesis, teaching, and service along with the scholarship of discovery, and that spoke to me as a regional campus faculty member.[11] Oral history linked my own research with teaching and connected the classroom to the community.

Precollegiate and college teachers assign oral histories in a variety of learning contexts, with everything from National History Day projects to courses on modern US history, language arts and writing, oral history methods, and a wide range of other topics from immigration history to World War II. Over time they develop strategies to disseminate the results of their work, such as high school teacher Linda Wood's student oral history publications including *What Did you Do in the War Grandma?*, middle-school teacher Michael Brooks' *Long, Long, Ago* magazine, and Indiana University Northwest history professor James Lane's *Steel Shavings* magazine.[12]

There are many ways to test the waters and to ease into interviewing assignments. The primary goal is, after all, student learning rather than producing an archival collection. I had previously used an interviewing assignment in writing courses at the University of Minnesota because I had concerns that more conventional writing assignments exploring students' own lived experiences lacked depth or analysis. I decided to change my approach and make an interview assignment that gave students ownership (or co-ownership) over substantive content that they could work with in class. Students used whatever recording equipment they could lay their hands on. A simple letter explained to narrators that this was a class assignment and that the student would share a copy of the recording and transcript with them. There was no plan to collect or archive the recordings. The transcript and the written profile based on it were the focus of this assignment, and the recording was secondary. Working with fourth graders on a family folklife project in Minneapolis, students carried home questionnaires and interviewed family members, writing in their versions of the answers or getting a little help in transcribing a story, song, or recipe. There was no attempt to record these interviews, either, and we subsequently compiled their work in a publication for the school community. In Middletown, a high school economics teacher and I asked her students to research the economic lives of women through oral history interviews. We published student-authored profiles on a website, and students hosted a reception for participants and community members featuring a slideshow about their work. In that project, few of the interviews were recorded and none were preserved, and yet students absorbed many of the benefits of a more formal oral history interview assignment.

Oral history work fulfills the expectations associated with constructionist, project-based, and inquiry-based learning—even more so when students choose their own research topics and plan the public programs. The experience is much richer when students engage with each other and with public historians and community members in determining the goals of the larger project, the questions to address, the selection of narrators, the evaluation of the interviews, and the analysis of data. In contrast to the didactic instruction of lectures, the scholarship of project-based and situated learning indicates that authentic problem solving provides a context in which learning is applied and retained to be used in new contexts later on.[13] However, given time constraints, it is often more practical to either let students choose their topic and narrator or to provide the class with a topical assignment and a list of narrators. With these limitations, oral history is similar to learning a craft that challenges students to move outside their comfort zone and requires autonomy and personal responsibility as well as collaborative work and decision making.[14]

While the assignment may vary, students need to learn the best practices in oral history, including respect for the narrator, background research, production of the best-quality audio recording possible under the circumstances, and the use of a formal release form along with the preservation of the recorded interview with a time-coded index or transcript to make the interview more accessible. Oral history guides emphasize the ethical responsibility to preserve materials. For example, the *Introduction to Oral History* published by the Baylor University Institute for Oral History advises that "Carefully preserved, the recordings carry the witness of the present into the future, where through creative programs and publications they can inform, instruct and inspire generations to come."[15] What comes after the interview, in terms of reflection, analysis, evaluation, preservation, and dissemination, is as integral to the learning experience as conducting the interview.

The analysis of texts—historical documents or literature as well as sound and visual evidence—is an integral part of teaching in the humanities. Recording oral history interviews themselves brings home to students the problematic nature of written and other forms of evidence as social and historical construct. Students come to understand the relationships of shared authority in the interview process itself, and class discussions following the interviews should explore the role of the student as interviewer and of the narrator as the source of memories, stories, meaning, and reflection. The narrator may have seized the opportunity to voice ideas or information that were not part of the initial research agenda, changing in the process the scope or direction of the research. In reviewing the interview experience, the student may find that leading questions or simply their own style led the narrator to be less forthcoming. Subsequent contact or correspondence extends the collaboration as the student seeks to clarify and expand on information. In reviewing the content, emotions, and perspectives expressed by the narrator, students may come to recognize how context, contingency, personalities, interests, and other forces shape both the past and the historical evidence. As project-based learning, the goal of sharing the interview with a wider audience enhances the students' responsibility to produce an accurate representation of the narrator's perspective and experiences.

In my own work teaching the US history survey, the assignment varied a great deal. As I became involved in the OHA, I placed greater emphasis on technical quality, release forms, preservation, and other best practices, but still with a fixation on the transcript as a primary source that both fostered student writing skills and provided an easily accessible resource for classroom discussion. Without a web-based platform like SoundCloud, which I will discuss later, sharing more than a handful of excerpts of recordings simply took up too much class time in rooms with 40 individuals. Students sometimes transcribed entire interviews that ran to 30 pages or more, while at other times they completed topical summaries keyed to counter numbers and transcribed 8 or 10 pages of the interview for class discussion. When students worked in pairs, transcribing or summarizing each other's interviews, they came away with detailed knowledge of two oral histories, and they could discuss the interview dynamics in greater detail. As part of the research assignment, students either annotated the interviews—adding their own comments on the process as well as historical context—or they wrote a description of the interview process, and a profile of the narrator contextualized the interview. My students generally selected their own narrators, often family members or older acquaintances, and carried away a resource that they and their families valued. Over time, the numerous interviews that students and narrators shared with the archives came to illuminate many dimensions of life in the Miami Valley.

The Miami Valley Cultural Heritage Project

During the 1990s, I taught US History and American Studies at Miami University's Middletown regional campus. Beginning in 1993, I had assigned students in the modern US history survey course to record and transcribe oral history interviews. Although not a requirement, students often came back with signed release forms, and I accumulated collections of interviews destined for the Miami University archives, 20 miles away in Oxford, Ohio. Concerned about keeping the interviews in the local community, I had turned first to the Middletown Public Library, but they were interested only in interviews recorded in Middletown, and my collections included narrators across the region between Dayton and Cincinnati, Ohio.

In a campus workshop during the 1994–1995 academic year, I learned to create hypertext markup language (HTML), the coding required to produce a basic website. The Internet offered the opportunity to share the interviews that my students and I had recorded back with narrators, families, and communities across the region. Funded by a small internal grant, I hired two students who were far more proficient in HTML, and together we produced the Miami Valley Cultural Heritage Project (MVCHP) website.

My goal was to produce a site that would engage ordinary people in the personal stories and diverse cultures of the region, known as the Miami Valley. The students, Mark Musselman and Adam Vary, taught me about both HTML and the Internet as we worked together. They persuasively argued that dozens of long transcripts would bore many visitors and that the long, unedited recordings would be difficult to store, access, and use. The audience I was aiming for would, we decided, want short, edited audio segments accompanied by transcripts linked together by keywords to an index or menu. Oral history excerpts published on the Center for History and New Media's companion site for the US History Survey use a similar format.[16] I harbored the longer-term goal of adding layers to the MVCHP so that interested visitors could access both the complete transcript and audio as well. My job was to select short, high quality, narrative excerpts to share online while the students coded the individual web pages and edited the recordings.

Visitors found streaming audio excerpts that ran from three to eight minutes embedded in web pages, along with a brief introduction to the narrator, and the transcript. Each excerpt was either a story or description of some facet of life of residents of the Miami Valley or their close kin, and the interview excerpts sometimes linked to longer transcripts. I also published a brief, practical guide to oral history interviewing along with a sample release form to encourage others to record family and community history.

The MVCHP used a very simple design with black text, orange headings, and burnt orange links on a white background. Speaker buttons played the audio segment while visitors read the transcript, and the index and keyword links led visitors to additional recordings on related topics such as labor history, military experience, Appalachian migration, African American migration, immigration, and ethnicity. A handful of photographs provided local context or illustrated the content. In addition to interview excerpts, there were student essays such as "Working in Steel: the Armco Employees Independent Federation"—the result of a collaborative field assignment in an introductory American Studies courses. For this project, students worked with Cityfolk, an organization then hosting the National Folk Festival, to document how workers decorated their hardhats and lunch boxes. This research, presented as part of the exhibit, was also available on the site.

We began this work well before I was aware of Cascading Style Sheets (first released in 1996), web development applications like Dreamweaver (developed by Macromedia in 1997), or content management systems (early versions were available in the late 1990s), so each page was individually coded in HTML. There was no way to track the number of visitors; Google Analytics came along in 2005.[17] Although I quickly shed the then ubiquitous blinking icon of a hardhat worker with a shovel, the site was under construction from its inception in 1996 to 2000 with a backlog of interviews to process and add before it would achieve the depth to be worth promoting.

The MVCHP lived briefly on the Miami University web server, and the oral histories, release forms, and transcripts live on in the Miami University archives. Grady Long, another student, processed the oral history collection in those archives. The close collaboration with undergraduates on this project and work with graduate students on the Oxford campus led me away from Middletown when I left to direct the Public History graduate program at Wright State University (WSU). Fearing cobwebs and dead links, and concerned about losing access to a site that named both students and narrators, I took down the MVCHP leaving behind the archival collection. I was fortunate that Mary Larson, then working on a much more ambitious digital production, took enough of an interest to review the project, and that the American Studies Crossroads' project asked me to write about it, or there would be no record the MVCHP had existed.[18]

Oral history interviews made for a challenging US history survey assignment on semesters at Miami University, but at Wright State University, that same course was compressed into two quarter-length courses that counted for still fewer credit hours. However, as I gave up assigning oral history interviews in the survey course, I began to teach undergraduate and graduate courses in oral history. While the assignment met the same pedagogical and curricular goals, learning about historical methods and the field of oral history took on new significance. Future teachers learned and critiqued both the process and the historical document, weighing the possibilities of using published interviews or assigning oral history interviews in their own classrooms. Public history graduate students needed to become proficient at the craft dimensions of oral history in order to work in archives and museums or to create documentary media productions.

Teaching Oral History in the Digital Age

When we launched the MVCHP, online oral history sites were proliferating and schools, archives, universities, and museums had begun sharing audio and/or video as well as transcripts. Today both college and precollegiate students are involved in producing oral histories as well as audio documentaries that are shared online via individual or class projects, ongoing school or university-based oral history initiatives, archives-based projects, thematic websites, and broader initiatives like the Library of Congress Veterans History Project. Students find opportunities beyond the campus as well. Two of my students participated in the Antioch College public radio station's Community Voices project, where they spent several Saturdays learning professional audio production techniques for broadcast and the web from local and national professionals including Neenah Ellis, Noah Adams, and Ira Glass.[19] Graduate student and coordinator of the WSU Student Technology Assistance Center, Will Davis, won a "Best Use of Sound" award from the Ohio Associated Press with his project.

The Internet has transformed opportunities for the dissemination of oral history recordings by students, scholars, and institutions, but it has also generated a host of questions and challenges for students, teachers, narrators, and archives. When I was still teaching at Miami University, the Center for History and New Media (CHNM) at George Mason University, the American Social History Project at the City University of New York, and the American Studies Crossroads Project hosted by Georgetown University were hubs of innovation that linked faculty interested in new media around the country.

The American Social History Project produced documentaries, supplemented by media-rich web resources as well as the interactive versions of the textbook, *Who Built America*, which introduced film and audio—including oral history interviews—into the classroom. Through the American Studies Association, I had the opportunity to work with Crossroads Project Director Randy Bass and many others on the 1998 publication, *Engines of Inquiry: A Practical Guide for Using Technology to Teach American Culture*, which featured a short article on the Miami Valley Cultural Heritage Project.[20] Randy Bass used thinking about new media as a springboard for thinking about teaching and learning as well as challenging definitions of scholarship. Founding director of the CHNM, Roy

Rosenzweig, wrote extensively, and often collaboratively, on the implications and potential of new media for teaching, scholarship, and public history. The center produced *History Matters: The U.S. Survey Course on the Web*, a web-based resource for teaching the US history survey that also offered up oral histories along with other primary sources edited for classroom use.

With the support of the Ohio Humanities, Rosenzweig came to Miami University in 1998, and we spent two days in fascinating conversations as we met with faculty in history, interdisciplinary studies, and computer sciences as well as with area teachers involved in National History Day. Rosenzweig pointed to the accessibility and porousness of digital resources and the democratization of scholarly work as it moved from print to the Internet. Bass and Rosenzweig brought the digital transformation sweeping through the humanities home to me, setting my own digital experiment in a wider context, introducing me to a network of academics and practitioners, and profoundly transforming my sense of myself as a teacher, scholar, and public historian.

Coming to WSU in 2000 and looking toward Dayton's 2003 Inventing Flight celebration, the graduate students and I found partners—the Montgomery County Historical Society, the National Afro American Museum and Cultural Center, and the WSU Special Collections and Archives—for a digital exhibit documenting Dayton in the Progressive Era. Funded by the NEH, the project made it through consultation and planning grants, with humanities scholars Tom Crouch, Judith Sealander, John Fleming, and Floyd Thomas, campus media professionals, and the staff of partner organizations. (When Montgomery County Historical Society, which held the core collections for the project, merged into a new historical organization with different leadership and priorities, they decided not to pursue the exhibit project.) Although not an oral history project, "Making Progress—Living and Working in Ohio's Miami Valley, 1890–1929" led us to mine local oral histories produced during and since the 1976 Bicentennial celebration as the public partners, graduate students, and I explored how best to represent the past on the Internet. In order to obtain funding, the best option at that time seemed to be to contract out the production of a customized, interactive, flash-media production by a web design company that had a record of nationally recognized productions. The consultation and planning grants had engaged students and me directly with both humanities scholars and campus based media producers. The leap to a costly contract with a production company presented a dilemma, because it separated the project from its pedagogical rationale as part of the public history program. The resulting production would have required further professional support to make significant changes or additions. This situation was changing just as our exhibit partnership fell apart.

My own experiences with oral history and with the Internet point to the crucial importance of treating public and digital humanities as a field of scholarship, encouraging faculty to engage in public work, and thus contributing to teaching and learning on campus as well as to the development of the humanities disciplines. Through their Digital Campus podcasts and then through THATCamps (The Humanities and Technology Camps or "unconferences"), the CHNM encouraged faculty and practitioners not only to adopt but also to adapt open source resources like the popular WordPress blogging platform. For-profit enterprises were marketing digital tools for course management, collections management, and research with varying price tags when CHNM launched customized, open source tools such as Zotero for research and Omeka for digital collections and online productions. CHNM developed these through a combination of grant funding, university support, and user communities. Serving on the Ohio Humanities board, I saw the potential of Omeka to reduce the costs of production for local history web projects like *Making Progress*, and as a result, more Ohio Humanities funds could support humanities scholars and public programs linked to web projects. We invited CHNM's then-managing director, Tom Scheinfeldt, to present Omeka to both the Ohio Humanities board and to Ohio public humanities organizations. With help from the CHNM staff, Ohio Humanities staff obtained NEH funding to develop a New Deal photography and oral history project with an Omeka-based online resource, and the Ohio Historical Society also launched an Omeka and WordPress site, *Civil War 150*. Learning from CHNM and Ohio Humanities, I quickly translated these experiences into experimental new media class projects and the development of a new course, History and New Media.[21]

While our Center for Teaching and Learning (CTL) enhanced my technological skills in respect to teaching in the classroom, the university library had also moved into the digital space with a focus on serving students. Its Student Technology Assistance Center (STAC) provided equipment to both digitize analog recordings and to digitally edit recordings. Today, the STAC staff teaches students the craft skills of digital media production one-on-one at high-end multimedia production workstations. The library partnership also provides new opportunities to share the students' oral history work online while avoiding the transience of the initial Miami Valley Cultural Heritage Project. The Special Collections and Archives—headed by Dawne Dewey, then a public history instructor—is the repository for student oral histories.[22] Archivist Jane Wildermuth became Head of Digital Services, responsible for planning, implementing, and supporting digital collections and services in the WSU Libraries. Her office partnered with other OhioLink institutions to create a system-wide digital repository with distinctive individual campus "portals" based in the libraries. This repository was hungry for content, including the student oral histories and the output of two Teaching American History grant projects. As a result, the library has come to provide training and tools, the archival repository, and the digital platform for student oral history interviews. With the support of the library, I no longer feel that I am a lone ranger off on some esoteric, challenging, and perhaps questionable pedagogical adventure.

This year, for the first time, I told students in an undergraduate Historical Methods course to create time-code summary indexes of their interviews and to transcribe only the portions that they planned to analyze and use in their writing—an approach that reaches back to Treleven's TAPE system. The day when an application will create a good draft transcript is not that far off. When it comes, it will supplement the audio or video that we will record, index, and disseminate today, and full transcripts will not be necessary. As an example, Wright State's Office of Veterans and Military Affairs launched a new project of veterans interviewing veterans, which is more accessible to students because they are freed from the hurdles, costs, and delays involved in transcription. While reviewing the recording to create the index or summary, students have the chance to assess their own work, identify significant content, and analyze narrative and performance.

By reducing the required transcription to a portion that is highly relevant to research, students still hone their writing skills, but they are freed from what is often an onerous process that does not contribute proportionately to learning goals in courses outside oral history methods. The result is that students have more time for analysis and the development of research papers or profiles. Michael Frisch, an innovator in accessing relevant excerpts of audio recordings that diminish the need for transcripts, has suggested that the machine-produced transcript has the potential to once again undermine the orality of oral history—the importance of listening to the recorded interview.[23] The persuasive arguments that Frisch and others have made for the centrality of the recorded interview should override the tendency to revert solely to the text for those serious about oral history research.[24] At the same time, new tools make it possible for students to share excerpts from their interviews quickly with one another. My graduate seminar in oral history used Audacity to edit oral history recordings and SoundCloud to share excerpts within days, enabling us to listen to excerpts online outside of course time and then discuss and review the excerpts in class.

SoundCloud works for sharing excerpts on the fly or developing a production to share with the public, but it does not replace the value of the archives and digital repository maintained by the library. The MVCHP and my move to Wright State University taught me the transience of faculty-led projects that are not grounded in an institutional context. At the same time, library platforms are also evolving. OhioLink has moved away from the shared digital repository that supported the WSU Libraries' Campus Online Repository [CORE] now called CORE Scholar. Led by Jane Wildermuth, the WSU libraries have turned to a platform developed at the University of California Berkeley, BePress, that is designed to share faculty, staff, and student publications—as well as work products—while also supporting formatting for digital and print publications. As Wildermuth's office worked to shift the oral histories and other collections to their new digital home, she and her staff helped me to find models for how to format future oral history collections online. In the future, we will break student interviews into five to eight minute segments (determined by content and flow). Students will prepare summaries for each of these segments and a master index that links to each segment. Visitors may access all the segments of an interview on a web page, or they may pull together excerpts linked by an index. (Unfortunately, BePress does not currently support either tagging or social tagging, which would permit visitors to create other ways to group and access excerpts.) In the past, when students used the full-length oral history recordings to produce web projects for the History and New Media course, the complete audio files were cumbersome to work with. These shorter segments are user friendly, allowing researchers and media producers to download and work with manageable files. We will share the transcripts that we do produce, linked to the media files and summaries. The ultimate goal is to facilitate the use of media files in student (and public) productions as well as to enhance access to and reliance on the media file rather than the transcript.

Oral History and Civic Engagement

In 2009, in another departure, I left the Department of History at Wright State University to join the Department of Urban Affairs and Geography, which, with its Center for Urban Public Affairs, had provided program management and evaluation support to previous grant projects. I keep a foot in the Department of History, teaching oral history, material culture, and topics in US social and cultural history, in addition to courses in community development and nonprofit administration in my new department. Funded by the Corporation for National and Community Service as part of a project directed by the American Association of State Colleges and Universities and five universities, I worked with the College of Education and Human Services and the Office of Service-Learning and Civic Engagement to develop a new minor in Youth and Community Engagement.[25] The community development course is taught in partnership with community-based organizations that determine, to a great extent, the scope and goals of student service-learning projects. In the Youth and Community Engagement minor, college students serve as coaches to K-12 students in developing initiatives to improve their neighborhoods and communities.

In 2012, Wright State University moved from quarters to semesters, providing more time for ideas to percolate and projects to take shape. Teaching courses in community development and nonprofit management with community partners, I am a part of a collaborative decision-making process regarding projects rather than the lone decision maker. The campus director of Service-Learning and Civic Engagement, our partners, and I determine how students' service can benefit both the students and the community. Oral history is new territory to many community partners, although it has a powerful role in community development, reflected in *Listening for a Change: Oral Testimony and Community Development* by Hugo Slim and Paul Thompson.[26] Neither advocates of service learning as pedagogy, nor these community partners, have oral history projects in mind when we sit down to plan, despite evidence that oral history is an effective tool for building social and cultural capital.

In Spring 2014, the community development students worked in three teams in Dayton's Westwood neighborhood. WSU students assisted with a series of senior forums to understand and share their perspectives on both past and contemporary life in the neighborhood. Through these forums, several seniors developed an agenda for change that they continue to address in weekly meetings. A second team conducted qualitative interviews with mothers of young children and daycare providers and then met with the Dayton Metro Libraries to advocate for an enclosed playground, to be attached to the children's section of a new library planned for the neighborhood. Additionally, several students attended and reported back on public listening sessions about youth and education. Challenged by a school administrator to integrate career-readiness and technical skills with their project, the third team worked with local middle-school students to conduct digital video interviews and record documentary photographs. While none of the students had worked on a documentary previously, the project culminated in the screening of their production, *Westwood: Then and Now*.[27]

By sharing responsibility with the Student Technology Assistance Center and ThinkTV Greater Dayton Public Television, students were able to access the training and digital tools to support this project. In collaboration with the WSU Libraries, we will also be able to preserve relevant interviews and documentation. Students and youth also learned attentive, critical listening, in different contexts, as a skill that they can carry forward. The interviews richly serve pedagogical goals, while enhancing intergenerational communications, supporting civic engagement in the schools, fostering community building, and contributing to the development of services that will improve the lives of the seniors. At the same time, these assignments brought home to me the challenge of learning both to listen and observe and to write about people, stories, interactions, and context. The attention to digital tools can distract from these crucial skills. The digital dimension of the project—from recording and editing processes to online repositories—is crucial, but it is no longer the centerpiece of the teaching and learning experience.

The Miami Valley Cultural Heritage Project reflected the digital turn in oral history and the humanities that took shape in the 1990s—an early and tentative effort both to use new media to share student work and to represent the narrative and performance qualities integral to the field. The transient life of the MVCHP is remedied in collaboration with archives and larger digital repositories. The more profound change is that students now have the tools to be the authors of their own productions, drawing on their own recordings or those produced by others. Ambitious projects linked by a topic or theme, such as the *Telling Their Stories Oral History Archives Project* developed by Howard Levin, director of Educational Innovation at Schools of the Sacred Heart in San Francisco, or the contributions that Bridget Federspiel's Dayton Public Schools high school students have made to the Library of Congress' Veterans History Project over the past decade are exciting practices.[28] At the same time, undergraduate and graduate students alike appreciate the opportunity to work individually or collaboratively on projects that may only last a semester, or might run as a thread through future academic work and into their lives beyond the university—like my own interviews with Lithuanians and Finns led to research on the role of the past in the social construction of identity that continues today. These diverse classroom-based digital oral history projects are linked by two values that are constants across my own teaching career: enhancing student learning through projects that require autonomy, craft, communication, and analysis while collaborating with students to engage with the community.

Notes

1. Truly democratic public servants, Jabbour and Fleischhauer took the time to discuss my project and helped me to set up appointments with the Folk Arts Office at NEA and the Division of Public Programs at NEH.
2. George Talbot, *At Home: Domestic Life in the Post-Centennial Era* (Madison, WI: Wisconsin Historical Society, 1976). The exhibit was on exhibition from spring 1976 through fall 1977 at the State Historical Society of Wisconsin with the support of the NEA. George A. Talbot also authored the exhibit catalog, an extended essay on amateur photography and the culture of American domestic life.
3. Dale E. Treleven, "Oral History, Audio Technology, and the TAPE System," *International Journal of Oral History*, 2(1) (1981): 26–45.
4. Chicago's Museum of Science and Industry used the LaserDisc technology in the *Newspaper* exhibit that opened in 1979. Carl Fleischhauer consulted on the *Six Generations Here* exhibit and videodisc project while on the staff at the Library of Congress. He then produced the optical videodisc for the American Folklife Center project, *Buckaroos in Paradise: Ranching Culture in Northern Nevada, 1945–1982*, that lives on as part of the *American Memory Project*: "The Buckaroos in Paradise Online Collection." *Library of Congress*, accessed May 1, 2014, http://memory.loc.gov/ammem/collections/buckaroos/index.html. Fleischhauer went on to direct the American Memory Project at the Library of Congress that distributed images, sound, and other files on LaserDisc before shifting to the Internet.
5. Funded by the NEH, my internship at Old World Wisconsin began a few weeks before the museum first opened in the summer of 1976. The book that I later produced from the exhibit (also funded by the NEH) and subsequent research—Marjorie McLellan, *Six Generations Here...A Farm Family Remembers* (Madison,

WI: Wisconsin Historical Society, 1996)—is the basis for an interpretive program "Life on the Farm," Wisconsin Historical Society, accessed May 1, 2014, http://oldworldwisconsin.wisconsinhistory.org/LifeOnTheFarm/.

6. The consultants' work examined immigrant and ethnic experience, working-class culture, and material culture and included: Kathleen Neils Conzen, *Immigrant Milwaukee, 1836–1860: Accommodation and Community in a Frontier City* (Cambridge, MA: Harvard University Press, 1976); Robert C. Ostergren, *A Community Transplanted: The Trans-Atlantic Experience of a Swedish Immigrant Settlement in the Upper Midwest, 1835–1915* (Madison, WI: University of Wisconsin Press, 1988); and Daniel T. Rodgers, *The Work Ethic in Industrial America, 1850–1920* (Chicago: University of Chicago Press, 1978). Kenneth Ames', *Death in the Dining Room and Other Tales of Victorian Culture* (Philadelphia: Temple University Press, 1995) was published later, but he brought a similar critical analysis to bear on the Krueger family photographs.
7. I am fortunate that my career was shaped by the Cooperstown Graduate Programs, part of the State University of New York at Oneonta, and the American Studies Program at the University of Minnesota, along with my experience at the Wisconsin Historical Society. The scholars that influenced my research at the time transcended disciplines, exploring oral narratives, nonliterate sources, narrative repertoire, performance, ritual, celebration, life course, and family from interdisciplinary perspectives. Their work included: Riv Ellen Prell-Foldes, *Prayer and Community: The Havurah in American Judaism* (Detroit, MI: Wayne State University Press, 1989); Elaine Tyler May, *Great Expectations: Marriage and Divorce in Post-Victorian America* (Chicago: University of Chicago Press, 1983); Yi-Fu Tuan, *Space and Place: the Perspective of Experience* (Minneapolis, MN: University of Minnesota Press, 1977); Gary Alan Fine, *With the Boys: Little League Baseball and Preadolescent Culture* (Chicago: University of Chicago Press, 1987); Allen Isaacman and Barbara Isaacman, *The Tradition of Resistance in Mozambique: The Zambesi Valley, 1850–1921* (Berkeley: University of California Press, 1976); John Modell, *Into One's Own: From Youth to Adulthood in the United States, 1920–1975* (Berkeley: University of California Press, 1989); Sara Evans, *Personal Politics: The Roots of Women's Liberation in the Civil Rights Movement and the New Left* (New York: Alfred Knopf, 1979); Juha Pentikäinen, *Oral Repertoire and World View: An Anthropological Study of Marina Takalo's Life History* (Folklore Fellows Communications, no. 219, Helsinki: Suomalainen Tiedeakatemia, 1978); Victor Turner, *Dramas, Fields, and Metaphors: Symbolic Action in Human Society* (Ithaca, NY: Cornell University Press, 1975) and *Celebration Studies in Festivity and Ritual* (Washington, DC: Smithsonian Institution Press, 1982); Barbara Myerhoff, *Number Our Days: A Triumph of Continuity and Culture Among Jewish Old People in an Urban Ghetto* (New York: Touchstone, 1980).
8. Marjorie L. McLellan, *Bread and Glue: Ritual and Celebration in Four American Families* (PhD dissertation, University of Minnesota, 1991).
9. In writing my dissertation, I also read widely in narrative research in folklore including Sandra K. Dolby Stahl, "A Literary Folkloristic Methodology for the Study of Meaning in Personal Narrative," *Journal of Folklife Research*, 22(1) (April 1985): 45–69.

10. I also learned from, and then taught, Charles T. Morrissey's interviewing methods, described in "The Two-Sentence Format as an Interviewing Technique in Oral History Fieldwork," *Oral History Review*, 15 (Spring 1987): 43–53.
11. Ernest L. Boyer's, *Scholarship Reconsidered: Priorities of the Professoriate* (Princeton: Carnegie Foundation for the Advancement of Teaching, 1997) was widely discussed but has only gradually been integrated into academia. The elective Carnegie Community Engagement Classification encourages institutions to re-examine and deepen their engagement with local, national, and international communities—a small step forward for public, digital, and oral historians teaching in universities.
12. Linda P. Wood along with staff and students from South Kingstown High School in Rhode Island, *What Did You Do in the War, Grandma?* (1989), *The Family in the Fifties, Hope, Fear, and Rock 'N Roll* (1993), and *The Whole World Was Watching, An Oral History of 1968* (1998). The latter project, produced in partnership with Brown University's Scholarly Technology Group with funding from the Rhode Island Committee for the Humanities, included both a publication and a website that featured transcripts keyed to real-time audio recordings of interviews. *What Did You During the War Grandma?* was redesigned in 1997 and is available online at http://cds.library.brown.edu/projects/WWII_Women/tocCS.html, accessed May 1, 2014. "Guide to The Family in the Fifties, Hope, Fear, and Rock'N Roll" in the University of Rhode Island Library Special Collections and Archives is available online at http://www.uri.edu/library/special_collections/registers/oral_histories/msg116.xml, accessed May 1, 2014. "Guide to The Whole World Was Watching, An Oral History of 1968," University of Rhode Island Library, Special Collections and Archives, accessed May 1, 2014, http://www.uri.edu/library/special_collections/registers/oral_histories/msg143.xml#relatedmatlink. Linda P. Woods and teacher Michael Brooks described their projects in the Organization of American Historians' *Magazine of History*, 11(3) (Spring 1997), ed. Clifford Kuhn and Marjorie McLellan, accessed May 1, 2014, http://magazine.oah.org/issues/113/. *Steel Shavings* magazine, produced since 1975 by faculty and students at Indiana University Northwest, examines the social history of the region through oral histories. Information about the collection can be found at http://www.iun.edu/~cra/steel_shavings/, accessed on May 1, 2014.
13. My understanding of project-based and situated learning comes, in part, from conversations over many years with Hilary McLellan, author of "Situated Learning: Multiple Perspectives," in *Situated Learning Perspectives*, ed. Hilary McLellan (Englewood Cliffs, NJ: Educational Technology Publications, 1996). There is a rich literature regarding these forms of constructivist learning including Laura Helle, Päivi Tynjälä, and Erkki, Olkinuora, "Project-Based Learning in Post-Secondary Education—Theory, Practice and Rubber Sling Shots," *Higher Education*, 51(2), (January 2006): 287–314.
14. Schyrlet Cameron and Carolyn Craig integrate the Library of Congress' Veterans History Project as project-based learning within the Common Core in *Project-Based Learning Tasks for Common Core Standards, Grades 6–8* (Quincy, IL: Mark Twain Media, 2014), 34–44.
15. Baylor, *Introduction to Oral History* (2014), accessed May 1, 2014, http://www.baylor.edu/content/services/document.php/43912.pdf. The commitment to preservation is also fundamental to the Oral History Association's *Principles and*

Best Practices, accessed May 1, 2014, http://www.oralhistory.org/about/principles-and-practices/.

16. Center for History and New Media, "History Matters: The U.S. Survey on the Web," accessed May 12, 2014, http://historymatters.gmu.edu/ offers up excerpts of oral history transcripts and audio. More recently Linda Shopes authored "Making Sense of Oral History" (2002), an interactive guide to the analysis of oral histories for teachers and students that I use in college classes. Accessed May 1, 2014, http://historymatters.gmu.edu/mse/oral/.
17. Håkon Wium Lie and Bert Bos, *Cascading Style Sheets, Designing for the Web*, 3rd edtion (Boston: Addison-Wesley Professional, 2005), xvii; Sean R. Nicholson, *Inside Dreamweaver UltraDev 4* (Corte Madera, CA: Waite Group Press, 2001), 8; Justin Cutroni, *Google Analytics* (Cambridge, MA: O'Reilly Media, 2010), ix.
18. Mary A. Larson, "Potential, Potential, Potential: The Marriage of Oral History and the World Wide Web," *The Journal of American History*, 88(2) (2001): 596–603; Randy Bass, ed., *Engines of Inquiry: A Practical Guide for Using Technology in Teaching American Culture* (Washington, DC: American Studies Association, American Studies Crossroads Project, 1997).
19. *Community Voices*, 91.3 WYSO, accessed May 1, 2014, http://wyso.org/community-voices. Neenah Ellis' tenure as station manager has brought innovative, community-oriented programming that engages audiences directly. In addition to *Community Voices*, WYSO has collaborated with filmmakers Julia Reichert and Steve Bognar to produce *ReInvention Stories*, profiles of residents who are reinventing their lives in the face of economic change, funded in part by the John D. and Catherine T. MacArthur Foundation and Wright State University.
20. Bass, *Engines of Inquiry*.
21. Public history students had worked with EAD (Encoded Archival Description) as well as Access, PastPerfect, and CONTENTdm to manage archival and museum collections. WSU has a centralized computer systems administration (CATS) that did not offer faculty access to a server for WordPress for several years and that still does not support Omeka. When they did turn their attention to requests for a blogging platform, we (CATS, the Center for Teaching and Learning, Communications and Marketing, and faculty like myself) invested over three years in evaluating and testing alternatives before launching http;//blogs.wright.edu/learn/. I had a great learning experience working with our computer systems, marketing, and student services staff, although the results were a long time in coming and have not engaged the wider university community to the extent that we had hoped. As a known computer geek, I was invited to serve on the committee—chaired by a Vice President—that adopted Drupal as the first content management system for the university website. That decision, by contrast, was made in a few months! In order to work with Omeka, I set up my own server account, although teaching with a personal account seemed to involve risks to student privacy as well the risk of getting hacked without the support of professional staff. Soon after, CHNM launched Omeka.net, a server platform for digital collections and exhibit projects like those my students and I were testing out with community organizations and museums.
22. Dewey and I worked together on a number of projects and public programs including the Cold War Technologies Oral History Project that she launched in

collaboration with the Wright-Patterson Air Force Base and other organizations. As a result, she hired Gino Pasi as an archivist for oral history. Dewey now ably directs both the public history program and the archives.

23. Michael Frisch, "From a Shared Authority to the Digital Kitchen and Back," in *Letting Go? Sharing Historical Authority in a User Generated World*, ed. Bill Adair, Benjamin Filene, and Laura Koloski (Philadelphia, PA: Left Coast Press, 2011), 126–137.
24. Ibid., 134.
25. I have served as WSU's interim director of Service-Learning and Civic Engagement since January 2014.
26. Soon after I began to see my work as oral history, I read extensively in Paul Thompson's work, including Hugo Slim and Paul Thompson, *Listening for a Change: Oral Testimony and Community Development* (Gabriola Island, BC: New Society Publishers, 1994). Also contributing to my confidence in oral history as an effective tool to foster civic engagement and local knowledge were: "Oral History and Community Involvement: The Baltimore Neighborhood Heritage Project" by Linda Shopes and "A Report on Doing History From Below: Brass Workers History Project," by Jeremy Brecher, both in *Presenting the Past: Essays on History and the Public*, ed. Susan Porter Benson, Steven Brier, and Roy Rosenzweig (Philadelphia, PA: Temple University Press, 1986), and Dolores Hayden, *Power of Place: Urban Landscapes as Public History* (Cambridge, MA: MIT Press, 1997). To validate the use of oral history in community development work, I also cite Elizabeth Duffrin's "Winning Over Naysayers," part of a case study of Richmond, Virginia's Fulton Hill neighborhood (New York: LISC Institute for Comprehensive Community Development, 2012), accessed May 23, 2014, http://www.institute ccd.org/news/4288. I have debated the validity of oral history as service-learning with numerous colleagues. My own article, "Case Studies in Oral History and Community Learning" highlights compelling examples of university-community partnerships—*Oral History Review*, 25(1–2) (1998): 81–112.
27. Will Davis, Coordinator of WSU's Student Technology Assistance Center, a high-end computing and media production center, skillfully guided my students as they worked with Westwood youth. Davis had previously enrolled in WYSO's Community Voices training and became the Community Voices instructor in 2014. This year, I directed his thesis project for a Master of Humanities in which he worked with youth at Dayton's Victoria Theater and Kettering's Fairmont High School on *Radio Waves*, accessed May 1, 2014, http://victoriatheatre.com/radio-waves. His own audio documentary about this project is at https://soundcloud.com/digitalhumanitarian/radio-waves, accessed May 11, 2014.
28. *Telling Their Stories*, accessed May 1, 2014, http://www.tellingstories.org/. Levin describes his work in, "Making History Come Alive," *Learning & Leading with Technology*, 31(3): 22–27, http://www.techlearning.com/techlearning/events/techforum06/HowardLevin_TellingStories.pdf and "Authentic Doing: Student Produced, Web-based Digital Video Oral Histories," *Oral History Review*, 38(1) (2011): 6–33. Bridget Federspiel was a graduate student in the Oral History course that I taught at Wright State University, in 2002–2003 and a fellow in the Dayton Teaching American History Project funded by the US Department of Education.

The work with her students can be found in a search of the Library of Congress Veterans History Project at http://lcweb2.loc.gov/diglib/vhp/html/search/search.html. Her work was recently highlighted in "Stivers student-teacher team win trip to Normandy," *Dayton Daily News*, March 26, 2014, accessed May 1, 2014, http://www.daytondailynews.com/news/lifestyles/stivers-student-teacher-team-win-trip-to-normandy/nfH4H/.

CHAPTER 6

Notes from the Field: Digital History and Oral History

Gerald Zahavi

Since the 1980s, technological changes—often collectively referred to as a *technological revolution*—have played an important part in the shaping and reshaping of traditional historical practices, expanding our capabilities to reach potentially vast audiences and to create innovative compositional works that meld visual, aural, and textual narratives into digital forms that can engage both scholars and the general public in discourse. The technological revolution has transformed how we collect, preserve, and disseminate historical knowledge. It has had an especially strong impact on oral history and on my own work as an oral, labor and business, and public historian.

Like others of the post-World War II, baby-boom generation, I was not born into an already underway digital revolution, but grew up with it through its embryonic and early developmental years. But unlike many others among my peers, an early appreciation (perhaps "infatuation" is a better word) with technology led me to throw myself wholeheartedly into the emerging revolution, totally immersing myself in its earliest conceptual and experimental stages. Proudly, in my small way, I tried to shape its development. So, let me start with some personal history which encapsulates indirectly the evolution of the relationship of oral history and digital technologies, and some of the ways in which oral history has been influenced by technical developments in the last three decades.

My embrace of computers, radio, sound recording, and digital technologies preceded my encounter with oral history. Back in White Plains High School in the late 1960s, I was already helping my nerdy friend build a computer in his home. At the same time, I was also developing an avid interest in radio, film, sound, and sound recording, spending hours listening to radio storytelling (mainly Jean Shepherd on WOR), an equal amount of time putting together shortwave radios and amplifiers (Heathkits), and getting heavily involved in my school's audio-visual center. I ended up as the student head of AV services as well as our theater program's sound crew. For our theatrical productions, I would record and edit sound effects, carefully splicing them onto 7" reel-to-reel tapes with ample white leader tape between segments so that I could identify and quickly cue up the next effect. I still have my cue tape from our production of *Our Town*. I also still vividly recall when we brought in blues guitarist Buddy Guy to play at our high school and how he kept asking me to crank up the volume on our sound system—and how difficult it was to do without generating feedback since he kept on moving the monitor speaker!

In high school, I learned to edit 16-mm film, repair tape recorders and projectors, and troubleshoot problems in all sorts of electronic devices. A considerable amount of my allowance and work money went into audio and camera equipment. (I worked every day after school and on Saturdays at a local pet store.) My prize possession from those years was my Tandberg reel-to-reel tape recorder, still working but now unused and in storage. It was also in high school that I began to explore *multimedia* composition. In 1968, instead of writing a traditional final paper in my English class, I asked for and got permission to produce a slide/sound presentation on Carl Sandburg. It was my first experience with script writing and my first multimedia production. I used several photographs from Edward J. Steichen, Sandburg's brother-in-law, in the production. No one taught me copyright law then. I could use anything—any image and any recordings of Sandburg—in my production. No problem. (Of course, who would ever sell and distribute my work?)

Sometime in my senior year, I also volunteered to work at WBAI in New York City, mainly editing tape. I had become a regular listener to the station for some time and wanted to get more involved in radio and especially in politically engaged broadcasting. Though a long illness in my final year of high school prevented me from pursuing my volunteer work for more than a short spell, my connection to WBAI and Pacifica (the foundation that owned WBAI and several other community, non-commercial stations around the nation) was nonetheless established. It would be a lifelong connection.

Once I got to college, my earlier projects with computers and my interest in technology and science made it almost inevitable for me to learn a couple of computer languages. I recall taking courses in Fortran IV and Basic. This was in the early 1970s, when I was still a science major taking lots of science and math classes, but by the time I graduated from Cornell University in 1973, I had switched to history and comparative literature. I had always been drawn to history and very much enjoyed both my high school and university history classes. The turbulent early 1970s and my growing interest in literature, anthropology, and especially intellectual history—the latter mainly through the influence of one of my teachers, Dominick LaCapra—made it increasingly difficult for me to imagine spending my life in a laboratory and unengaged with the world. It was also at Cornell that I obtained my first experience with oral history, working with a friend on a series of interviews conducted in Buffalo, New York, focusing on the Attica prison uprising of 1971, and used in an article for the *Cornell Daily Sun*. (The recordings were sent down to WBAI for their use and have since been lost.) I was still using an analog reel-to-reel recorder—this time a portable SONY recorder that weighed around ten pounds.

After a short spell in Europe and England, and taking odd jobs in upstate New York, I entered graduate school in the mid-1970s, receiving my MA in 1978 and my PhD in 1983, both from the Maxwell School at Syracuse University. Though my MA was in European history, I switched to a US focus for my PhD. My research while I was a European history student had focused on the intellectual history of critics of capitalism—particularly English, pre-Marxian socialists. When I switched to US history for my PhD, I retained my interest in capitalism and its critics but concentrated less on intellectual history and more on social, labor, and business history. Starting around 1978, I was drawn to the study of corporate welfare capitalism. I had become interested in questions pertaining to the response of workers to corporate paternalism and decided to focus my dissertation work on one company renowned for its corporate welfare work—the Endicott Johnson Shoe Company based near Binghamton, New York.

Though poorly trained in oral history at the time, I adopted it quickly in my dissertation research, in part because it was absolutely necessary for the sort of analysis I had in mind—one balancing the voices of managers and workers and involving a close, highly detailed, dialectical analysis of complex economic institutions, relationships, and phenomena that required going beyond textual, archival documents. My examination of welfare capitalism through an analysis of one firm uncovered how workers and managers jointly created and sustained a business culture founded on mutual worker and management obligations. Corporate paternalism at Endicott Johnson and at other firms was built precisely on *negotiated loyalty*—a loyalty preconditioned on escalating labor expectations and fulfillment of managerial promises. Oral history provided me with a methodology to gauge the response of workers to corporate paternalism through their own voices and memories.

While I was a neophyte to oral history, I was far more familiar with its tools. I took my Sony reel-to-reel recorder with me on my regular trips down to Endicott, Johnson City, and Binghamton, New York, to conduct my interviews but soon replaced it with a high-quality cassette recorder. At the time, analog open-tape recording was still the standard for high-quality recording, though I would hardly call my recordings at the time *high quality*. I was much more concerned with obtaining an audio record of my exchanges with Endicott Johnson workers and managers for use in my dissertation than I was in obtaining radio-quality recordings (which was too bad, as I had to employ some audio artisanship recently to clean up the interviews well enough for Joe Richman to use in his National Public Radio piece on George F. Johnson and corporate paternalism at Endicott Johnson).[1]

My research fieldwork soon led me to embrace oral history wholeheartedly. I loved the personal relationships it helped me establish, the depth of insight it provided me into daily work relations in a corporation, and the intellectual labor required to make sense of the rich and diverse personal perspectives and memories that I collected. While my utilization of oral history was not very sophisticated initially, the influence of such writers as Alessandro Portelli, Studs Terkel, Ron Grele, and Michael Frisch in the late 1980s and early 1990s helped me refine my interpretive skills. When I received a tenure track position at the University at Albany, SUNY, it was in part because of my oral history and computing expertise. The job description, if I recall correctly, placed a premium on such skills. I was expected to develop courses in local and regional history and oral history to serve the public history program that the department had recently created, as well as a course in quantitative methods for historians. I soon became the department's "tech person."

In 1985, the year I got my job, there were only two of us with personal computers—myself, with a brand new Zenith Z-series computer with an 8088 processor, and Robert Dykstra, who was working on a highly quantitative study of race relations in nineteenth century Iowa. Most serious computing at that time still relied on terminals linked to mainframes in centralized computing centers, with data saved on tape and IBM cards. In fact, when I developed a quantitative history course soon after my arrival, data analysis was carried on in that way, with the mainframe processing submitted data. Students would submit their *jobs* via a terminal and then pick up their SPSS and SAS (statistical data analysis programs) output at the university computing center. Within a few years, we shifted to a PC-based system, but it was hardly as simple or elegant as present-day personal computers. By the early 1990s, I had also begun working on the Internet, joining H-NET (begun in 1992 by Prof. Richard Jensen at the University of Illinois at Chicago) very early on and developing, with the assistance of one of my graduate students, one of the earliest departmental websites on our campus. That website soon became a vehicle for delivering all sorts of multimedia content to viewers.

With the expansion of digital technologies in the 1990s—particularly digital audio in the form of DAT (Digital Audio Tape) and MiniDisc recording—I soon began shifting my oral history recordings to digital format (always being careful to create analog cassette copies, since DAT tapes were very poor storage mediums). By that time, I was also teaching an oral history course, and though the department had invested in cassette recorders which I continued to utilize in my course, as funds became available, I began to slowly update our loaner recording kits to digital formats (DAT and MiniDisc), and later, the more reliable, cheaper, and easier to use SD/SDHC card recorders.

At the same time, as recording technologies were evolving, a number of companies and technical consortiums had been developing media compression standards and formats that would allow for the delivery of media over the Internet. In the mid-1990s, RealNetworks introduced an audio-streaming format that could deliver audio over low and high bandwidth networks, and they soon followed up with video streaming. The MP3 format was also evolving through the early 1990s. By the middle of the decade, it was possible to see a future in which both audio and video content could be integrated and easily delivered to users around the world. Excited about all of these prospects and deeply believing that all historians should be *public* historians, sharing their work with as broad an audience as possible, I began putting recordings on our servers and making them available to Internet users as streaming files, and soon as both streaming and non-streaming formats (MP3).

Public oral history projects became an important focus of my academic work and reflected my concern with the overly inward-focused and arcane debates taking place in my own and related fields. Of course, there were others who were equally concerned about this. Sherna Gluck was beginning her pioneering work at California State University, Long Beach, creating the Virtual Oral/Aural History Archive, which emphasized the importance of providing access to "the actual spoken words of oral history narrators, rather than seeing a written version of them in the form of a transcript."[2] Other oral history programs soon followed her lead. At the University at Albany, SUNY, we began a number of community-based oral/visual history projects whose aim was to record and disseminate interviews and documentary work over the World Wide Web (WWW); most are still in progress. "The Glovers of Fulton County, New York" was a research and documentation project created to explore and record the history of a once major but then near-death industry in upstate New York—the glove industry of Fulton County. It yielded dozens of hours of videotaped oral interviews, thousands of scanned documents (government reports, union publications, newspapers, memoranda), and a large collection of material objects related to the glove industry. The material is currently archived and in storage, but we hope to mount a multimedia exhibit and complete a documentary film utilizing these materials in the near future. A preliminary website was created for the project during the latter part of the 1990s, and included some of the earliest examples of online video interviews. Now, of course, such interviews are widely available and common practice; in the 1990s they were not so.

Other projects were also begun in the 1990s and early 2000s: "The General Electric Corporation: A Digital History," "Life and Labor in a Corporate Community: An On-Line History of the Endicott Johnson Corporation," and "Capital Voices, Capital Soundscapes." The latter yielded a number of student-produced, oral history-rich documentary radio productions focusing on topics related to the history of the Capital Region of New York. Susan McCormick, who became involved in much of this early work after the mid-1990s, and I also began to offer talks and workshops on oral history and digital history in Fulton County, Albany, and at historical conferences. In 2000, we ran a workshop on "Oral History as Public History" devoted to technical instruction in multimedia oral history presentations (Oral History Association Meeting, 2000), and I delivered presentations on history and digital multimedia scholarship at Skidmore College, Cortland College, Oneonta College, and at the American Historical Association and Oral History Association annual meetings.

Oral history archival projects became an early emphasis of my work as a labor historian in the 1990s and early 2000s as well—again, with an emphasis on utilizing digital technologies and the WWW to make historical resources available to a wide audience. Working with Jane LaTour, we created an online audio and manuscript archive focusing on the work of Nathan Spero in the United Electrical Workers Union (UE): "Nathan Spero: A Life in the UE." Spero served for four decades as the research director for the United Electrical Workers Union, from 1944 until 1983, and before that, he worked as statistician for the National Research Project on Productivity and Technological Change for the US Department of Labor (from 1937 until 1943). After digitizing the interviews conducted by Jane LaTour between 1994 and 1996 as well as supplementary documentary materials on Spero's life and career, LaTour and I—with the assistance of Patricia Logan, Aaron Wunderlich, and Susan McCormick—put together the oral history-rich site devoted to Spero and his career. Continuing similar work, more recently with the help of one of my graduate students, Carolyn Wavrin, we also worked on an oral history online archive to supplement LaTour's book, *Sisters in the Brotherhoods: Working Women Organizing for Equality in New York City* (2008), which focused on pioneering young women who sought to break down gender barriers in traditionally male, blue-color jobs in New York City during the 1970s and 1980s. The archive documented LaTour's 20-year long oral history efforts to document the stories of these women and featured audio and transcripts of the Sisters interviews as well as material from LaTour's book.[3]

I began other labor history related oral (and *aural*) history projects, including the "Sam Darcy Oral History Project," a collection of interviews, oral memoirs, and home recordings pertaining to the life and career of Communist Party USA leader and activist Sam A. Darcy. All of these became part of what I called the "U.S. Labor and Industrial History World Wide Web Audio Archive" and linked to *Talking History*, a production, distribution, and instructional center for all forms of "aural" history that I began in the mid-1990s.[4]

The widening of digital horizons in the mid-1990s and my reawakened interest in radio in the early 1990s led me into two enterprises that had widespread implications. In 1996, consistent with my long-term interest in the use of both old and new media to communicate history to a wide audience, I founded *Talking History*, an aural history production center with a weekly FM radio program that is also broadcast over the Internet (www.talkinghistory.org). A year later, with Julian Zelizer and Susan McCormick, I co-founded the *Journal for MultiMedia History*. By that time, McCormick had also joined me as co-producer on *Talking History*.

I had begun broadcasting on WRPI-Troy in 1995, producing a politically engaged show called *Capital District Progressive Radio*, and soon afterward, I tried to create a segment devoted to history. Taking a name suggested by my colleague Robert Dykstra, *Talking History* began as a local history show but soon adopted a broader focus. Partnering with Creighton University at first, and later independent producers and stations around the country and world (including broadcasting agreements with the Australian Broadcasting Corporation, Radio Netherlands, and the Canadian Broadcasting Corporation), we began showcasing history-focused radio programming on the show. Typical broadcasts included documentaries, interviews, oral history segments, and archival audio segments. By the late 1990s, we began preparing streaming versions of our broadcasts in low-bandwidth versions, available to WWW listeners equipped with 300+ baud modems, and soon afterward in 2002, we moved into higher bandwidth online content, including MP3 format. We also provided an early storage platform for a number of independent radio producers, digitizing and placing their radio documentaries on our servers for delivery through their own websites. Both David Isay and Joe Richman took advantage of our services; now their prominence and more stable financial situations enable them to archive their programs on their own sites and on National Public Radio's servers. David Cohen (New Jersey Historical Commission), Dan Collison (Long Haul Productions), and oral historian Charles Hardy III also contributed their radio productions to *Talking History*, and their work is still available on our site.[5]

Talking History's aim was to "provide teachers, students, researchers and the general public with as broad and outstanding a collection of audio documentaries, speeches, debates, oral histories, conference sessions, commentaries, archival audio sources, and other aural history resources as is available anywhere." It sought to expand the aural focus of historical research and scholarship by providing a production, distribution, and pedagogical venue for aural history (including *oral* history) and encouraging scholars and documentary producers to explore the audio dimensions of our past. Pedagogically, my work on *Talking History* was soon tied to a new course I created in the History Department titled "Historical

Documentary Production for Broadcast and Internet Radio," and it soon caught the attention of the *Chronicle of Higher Education* and National Public Radio producers. The course, then probably the only one focusing on narrative aural history production taught in a history department, focused on the historical study of sound, soundscapes, and sound recordings, oral history techniques for radio, aural history composition techniques (radio documentaries, features, aural essays, and museum audio installations), and audio delivery technologies to communicate historical ideas to broad audiences. It included coverage of textual and archival audio source research; twentieth and twenty-first century historical radio documentary work; analysis of audio documentary forms and non-fiction storytelling techniques; script writing; technical instruction in the art of audio recording and postproduction editing and mixing; discussions of audio preservation and restoration techniques; and an introduction to traditional and modern technologies for the transmission and dissemination of documentary and related audio work. The *Chronicle* recognized it for its unique pedagogical approach and organization, its unique exploitation of the WWW, and its involvement of guest documentary producers—many independent radio producers then engaged in historical documentary production, including David Isay, Dan Collison, and Joe Richman.[6]

By the late 1990s, oral history and digital history had become a central aspect of all of my research projects, but I also recognized that while the tools and the technology of the field had begun to shift toward a digital standard, the fundamentals had not really changed: the dialogic nature of oral history, the need to establish good rapport with an interviewee, the importance of deep listening, the critical role of patience and good interrogatory skills. Neither digital recording nor delivery innovations altered these, and they remained central to my pedagogical work. I emphasized to students, who seemed to want to focus on the tools of the trade, that recording technologies come and go but the essential aspects of good oral history, on the level of the individual interview, are not matters of technology.

Through the 1990s and into the following decades, the proliferation of affordable digital recorders made extremely high quality recording possible and accessible to a growing number of people. Perhaps most exciting were the new compositional opportunities and possibilities that opened up for oral historians. Interpretive essays, if published in digital form, could now integrate multimedia content and oral history excerpts directly into analyses of that content. This became a major element in my initial venture into digital online publishing: *The Journal for MultiMedia History* (*JMMH*). Initiated in 1997 and co-edited by Julian Zelizer and me with Susan McCormick as the Managing Editor, the *JMMH* was established as a free, peer-reviewed, online historical journal devoted to presenting hypermedia articles, documentaries, aural essays, and reviews. At the time, the idea of publishing historical multimedia essays and content as electronic journal *articles* was still new, and there certainly was no peer-review process in place to evaluate such scholarship. While a handful of text online journals were starting to publish around the time we began—for example, *The Journal of the Association for History and Computing* (begun in 1998) and *Common-place* (begun two years later, in 2000)—no historical journal was devoted to exploring the potential of *true* digital publishing, where audio, video, graphics, and text could be integrated into intermedia compositional works. The *JMMH* was to demonstrate how hypertext and multimedia technologies could truly transform research, documentation, and dissemination of historical scholarship. With respect to oral history, the World Wide Web had introduced a revolution in publishing. It made the dissemination of both raw oral history interviews (in transcript and audio versions) and interpretive, analytical essays that relied on oral history easier and more ambitious. Now online projects could reach hundreds of millions of potential viewers and readers instead of a few thousand.

The *JMMH* came out of conversations taking place within the History Department by members of a newly formed History and Media Committee, which I chaired. The participants were devoted to broadening the margins of academic historical discourse and pushing academic history into the realm of both traditional and digital media scholarship and publishing (audio/radio, video/film, television, CD/DVD, Internet/WWW). As we explained in our first issue, "We wanted to bring serious historical scholarship and pedagogy under the scrutiny of amateurs and professionals alike, to utilize the promise of digital technologies to expand history's boundaries, merge its forms, and promote and legitimate innovations in teaching and research that we saw emerging all around us."[7] We especially believed, then and now, that digital, hypermedia publication would be more exciting and accessible to an entire universe of non-academic readers and would inject history more directly into public discourse.

The first issue of the *JMMH* came out in 1998. It was a visually rough but promising beginning and demonstrated in its first article—"The 1939 Dairy Farmers Union Milk Strike in Heuvelton and Canton, New York" by Thomas J. Kriger[8]—how oral history can be incorporated into digital compositions. Kriger's article included audio excerpts from several oral history interviews conducted in the course of his research of the 1939 Dairy Farmers Union strike. Other pieces in our first issue included recordings of speeches as well as multimedia reviews of CD/DVDs, websites, films, and radio programs, all integrating audio, graphic, and video elements into the reviews. The *Journal* was clearly promoting a digital, mixed media (*hypermedia* or *intermedia*) approach to historical argumentation and evidentiary presentation, and it began to challenge traditional approaches to historical publication.

Of course, none of what we wanted to accomplish was easy, and we faced many challenges. Our first was obtaining quality submissions. We were *not* looking for traditional text submissions. We were looking for *articles* that fully utilized the potential of digital presentation and argument and that focused on visually and aurally rich topics (especially those that were directly experimenting with new hypermedia grammars and digital compositional ideas). In many ways, we had to become missionaries as well as very actively engaged editors. We held long and productive discussions with potential authors, convincing them of the merits of presenting and exploring their subjects in hypermedia format. We often worked tirelessly with authors to develop an effective structure and form for their submitted pieces. Editorial work at the *JMMH* thus clearly involved more than the traditional work of journal editors. We also faced the daunting challenge of producing for multiple operating systems, browsers, and monitor screen sizes and resolutions, which at the time was a far greater challenge than today, when more uniform coding standards have been adopted and widely disseminated.

There were other challenges, as well. The journal was a path-breaking venture and few historians then had the necessary skills to engage in multimedia publishing. (Most still lack those skills today.) Production of core articles and reviews involved preparing the visual and aural elements as well as writing much of the HTML and JavaScript code that created the look of the finished pieces. Authors submitted essays and the elements that editors had to incorporate into the finished online articles. In a few cases, submissions did arrive in fairly polished hypermedia form. Tom Dublin and Melissa Doak submitted an excellent and generally precoded piece titled "Miner's Son, Miners' Photographer: The Life and Work of George Harvan," an in-depth profile of documentary Pennsylvania photographer George Harvan and his work. Though additional coding work was required by *Journal* editors, the bulk of the most difficult coding was done by Dublin, Doak, and their technical consultants at the University of Binghamton, SUNY. The final work included 280 photographs, hours of oral history interviews, flash slide exhibits, and an analytical essay with hyperlinks to various visual and aural resources. That piece demonstrated how to utilize the full potential of electronic publishing in scholarship focusing on photographic, video, or audio subject matter.[9]

Charles Hardy and Alessandro Portelli's "'I Can Almost See the Lights of Home': A Field Trip to Harlan County, Kentucky," was another important project that appeared in the *JMMH*, and it challenged traditional publication conventions more directly—especially those followed by oral historians.[10] Although begun as a purely audio project, Hardy and Portelli decided after we spoke to submit it to the *Journal* and reconfigure it for the World Wide Web. After outside reviewers recommended publication, *JMMH* editors worked closely with the two authors to construct the online version. "I Can Almost See the Lights of Home" suggested a new mode of thinking about and presenting oral history. Joint authors Portelli and Hardy termed their work an *aural essay* and—borrowing freely from my introduction to the essay in the *JMMH*—attempted to create a new aural history genre that counterpoised the voices of subject and scholar in dialogue, the dialogue that took place in the real time of an oral interview *and* the one that followed as interpretations and scholarship were generated. Dialogic elements pervaded the work in both the conversations between Portelli and Harlan County residents and in the verbal exchanges between Portelli and Hardy as they discussed the underlying themes and meanings imbedded in the interviews. "I Can Almost See the Lights of Home" was also clearly an intellectual manifesto that both challenged *oral* historians to engage the "orality" of their oral sources and encouraged historians to consider alternative modes of presenting interpretations (modes that would make the very act of interpretation transparent).

With such works as "Miner's Son, Miners' Photographer: The Life and Work of George Harvan" and "I Can Almost See the Lights of Home," the *JMMH* had a profound influence on promoting original works that explored new presentation modes for oral history. As a review article in the *Journal of American History* acknowledged, "The effect of the sound files paired with contemplative writing is more evocative of place and thought than are most standard journal articles."[11] That's precisely what we hoped multimedia publishing would help accomplish.

Still, in spite of its promise, the *Journal* faced formidable obstacles beyond those already mentioned. It operated on a very small budget and with a volunteer staff, and though it clearly improved from one issue to another—visually, conceptually, and stylistically—and matured in its understanding of the syntax of digital scholarship, it could not sustain that evolution without considerable resources and full-time editorial attention. Furthermore, soliciting and attracting quality submissions continued to be a major challenge. Younger historians, very much concerned about professional visibility and enhancing their academic credentials and reputations, remained hesitant to devote considerable time to digital projects that might or might not be viewed as equal to traditional text publications. These and other obstacles limited the progress of the *JMMH* and led to the cessation of publication after three issues. Attempts to locate another home for the *Journal* continue, and Matrix[12] at Michigan State University has expressed interest in reviving the journal. Only time will tell if it will indeed be resuscitated.

My work in multimedia digital scholarship and on the *JMMH* led me into several projects that sought to more directly explore hypermedia composition. One experiment in hypermedia writing, which I began over a decade ago and which continues to evolve, focused on General Electric's Association Island. The essay, containing aural and visual elements as well as extensive oral history excerpts woven around an analytical, textual narrative spine, examines a half-century of managerial recreational life at a corporate-sponsored retreat on Lake Ontario. While others were experimenting with *digital non-linear narratives*, facilitated by dense hyperlink composition, I was more engaged with *multimedia narrative writing*. I wanted to preserve my authorial voice and interpretive responsibility but provide readers with more direct access to my evidence and a richer and deeper appreciation of a subject than traditional textual narratives could offer. The media components—corporate films, oral history excerpts, archival recordings, music and songs, photographs and graphics—contributed to the overall interpretation of the transformation of General Electric's corporate culture over time and suggest how multimedia elements can be combined into a single argument presented in digital hypermedia form.

A second project, *The Lamont-Doherty Earth Observatory of Columbia University: An Oral History, 1949–1999*, was released as a double audio CD set. Produced as an anniversary retrospective on the history of one of the nation's top oceanographic laboratories, the compilation demonstrated as early as 1999 how online technologies could facilitate the collaborative work of oral historians.[13] The final product was produced by the Columbia Oral History Research Office (with then-director Ronald J. Grele), Lamont-Doherty Earth Observatory staff, and me and a gifted undergraduate student at the University at Albany. The process of putting together the two-CD compilation involved digitally processing and weaving together narrative introductions with oral history excerpts (from interviews conducted by historian Ron Doel) into a coherent sequence of narrative interviews highlighting some of the central achievements of the observatory. I served as the producer and audio editor/engineer of the project, and the work of composing the finished CDs was carried out through an innovative, collaborative, online review process that I implemented at the University at Albany. Each edit was uploaded in streaming audio format for the project participants to listen to and comment on. Suggestions for revisions were shared through email exchanges, and revisions were made and reposted. The process continued until each segment was satisfactory to the members of the editorial board. Now, in the second decade of the twenty-first century, collaborative online work is easy and common. This was not the case in the 1990s; the ease of working collaboratively on digital oral history projects was greatly facilitated by the development of digital recording, encoding, and transmission technologies.

Perhaps the greatest impact on *oral* history is the development and maturation of visual oral history. My current work continues to heavily utilize oral history—but in the form of videotaped interviews—and it suggests the changes that are currently underway in the field. Major oral history programs are turning more and more to the use of digital video cameras to document interviews and to conduct field recordings in non-traditional outdoor and industrial settings. The growing accessibility and affordability of digital video cameras, together with the recognition by many oral historians that visual interviewing may offer important supplemental and substantive information for oral historians, have led many into the realm of video history, including DVD and online distribution of video interviews. The growing presence of documentary film sessions at the Oral History Association also suggests that visual documentary scholarship is indeed growing in importance to the organization and the field in general.

Technical advances will undoubtedly continue to expand oral historians' abilities to share their work with wider and larger audiences and to engage in innovative compositional work that heavily incorporates oral history recordings. Visual oral history and online documentary work will no doubt expand. The ability to conduct oral histories (and visual interviews) online will inevitably introduce controversies about distance interviewing and its impact on rapport and trust, as well as on how such practices impact the quality of dialogic exchange. The ease of digital video recording of interviews and growing expectations for *all* interviews to be video recorded will challenge our emphasis on *orality*, as visual oral history introduces another layer of content to be recorded and analyzed by scholars. All of these layers will make the field that much more exciting in the future, but as I noted earlier, the central concerns of oral historians will remain the same: building a trusting space for dialogue, generating probing and effective questions, developing deep listening skills, and producing sophisticated interpretations of orally expressed memories.

Notes

1. Joe Richman radio documentary, "The Legacy of George F. Johnson and the Square Deal," *Radio Diaries*, accessed May 5, 2014, http://www.npr.org/2010/12/01/131725100/the-legacy-of-george-f-johnson-and-the-square-deal.
2. *The Virtual Oral/Aural History Archive*, "Welcome to the Virtual Oral/Aural History Archive," accessed September 8, 2012, http://www.csulb.edu/voaha.
3. *Sisters in the Brotherhoods: Working Women Organizing for Equality in New York City*, accessed September 8, 2012, http://www.*talkinghistory*.org/sisters/.
4. *US Labor and Industrial History World Wide Web Audio Archive* (Department of History, University at Albany, State University of New York), accessed September 8, 2012, http://www.albany.edu/history/LaborAudio.
5. *Talking History*, "Contributing Producers," accessed September 8, 2012, http://www.talkinghistory.org/producers.html.
6. *The Chronicle of Higher Education*, 47(46) (July 27, 2001), accessed September 5, 2014, https://chronicle.com/article/At-SUNY-Albany-History-Stu/3094/.
7. *JMMH*, 1(1) (1998), accessed June 5, 2014, http://www.albany.edu/jmmh/vol1no1/v1n1.html.
8. Thomas J. Kriger, "The 1939 Dairy Farmers Union Milk Strike in Heuvelton and Canton, New York: The Story in Words and Pictures," *JMMH*, 1(1) (1998), accessed June 5, 2014, http://www.albany.edu/jmmh/vol1no1/dairy1.html.
9. *JMMH*, 3 (2000), accessed June 5, 2014, http://www.albany.edu/jmmh/vol3/harvan/index.html.
10. Charles Hardy III and Alessandro Portelli, "'I Can Almost See the Lights of Home': A Field Trip to Harlan County, Kentucky," *JMMH*, 2(1) (1999), accessed June 5, 2014, http://www.albany.edu/jmmh/vol2no1/v2.html.

11. Mary Larson, "Potential, Potential, Potential: The Marriage of Oral History and the World Wide Web," *Journal of American History*, 88(2) (September 2001): 600–601.
12. MATRIX Center for Digital Humanities & Social Sciences, accessed June 5, 2014, http://www2.matrix.msu.edu/about.
13. *Lamont-Doherty Earth Observatory of Columbia University: An Oral History, 1949–1999* (New York: Columbia University Oral History Research Office, 1999).

CHAPTER 7

Densho: The Japanese American Legacy Project

Tom Ikeda

Densho: The Japanese American Legacy Project is a nonprofit organization based in Seattle, with ten employees and an annual budget of $1 million dollars.[1] Over the last 18 years, it has produced a website that has emerged as an important source of learning and teaching about the World War II incarceration of 120,000 Japanese Americans. Within the course of the past year, hundreds of thousands of people from around the world have visited Densho's onsite offerings, including its still-expanding collection of 800 video interviews with Japanese Americans who were incarcerated in wartime American detention facilities.

Densho was a pioneering organization in its use of the Internet to make archival content freely available. This chapter examines some of the steps that led to the methods and tools used by the project to collect, preserve, and share its materials, along with examples of how people have used Densho's online resources.

Origins

In 1995, a planning group of 20 Seattle-area volunteers, almost all Japanese Americans, met to discuss doing a Japanese American oral history project. They named the group *Densho*, a Japanese term meaning, "to pass stories to the next generation." In the early meetings, a simple question helped define the purpose of the project. "What would happen if Japanese Americans who were unjustly incarcerated during World War II shared their stories?" This question shifted the discussion from simply conducting and preserving oral history interviews to a much broader discussion of how the stories could be used to promote and advance the country's democracy. Densho understood how war hysteria and racism led to the mass exile and unjust imprisonment of 120,000 Japanese residents and Japanese American citizens. These innocent individuals were forced from their West Coast homes without trials or hearings and placed in tar-papered shacks surrounded by barbed wire and guarded by armed soldiers. Everyone who was part of Densho was either incarcerated during World War II or had relatives or close friends who were.

Although World War II had ended 50 years earlier and race relations had improved since the 1940s, many in the planning group in 1995 believed that similar racial or religious profiling could happen again if the country was threatened. Densho decided that it would be important to design a project that did more than preserve the Japanese American incarceration story. The project also needed to share the stories in a way that kept them alive so that similar targeting would not happen to another group. In particular, it would be valuable for teachers and students as they were learning about democratic ideals and the Constitution to have access to these first-person stories of formerly incarcerated Japanese Americans. The group quickly realized that public access to these personal stories would be a requirement, but how should these stories be made available?

Fortunately, Densho had the opportunity to learn from the Survivors of the Shoah Visual History Foundation, a Los Angeles-based organization that was started by filmmaker Steven Spielberg after he completed the film, *Schindler's List* in 1993. In 1995, the Shoah Foundation was in the midst of a huge project to conduct, preserve, and organize tens of thousands of video oral history interviews with Holocaust survivors. A Densho team—led by Scott Oki, a former Microsoft executive, and Tom Ikeda, author of this chapter and a former Microsoft general manager in the Multimedia Publishing Group—visited the Shoah project to see if they could produce something similar.

In trailer offices on the back lots of Universal Studios in Los Angeles, the Seattle group was introduced to a massive program to conduct and process 50,000 interviews with survivors of the Holocaust. The well-funded Shoah Foundation project was very sophisticated. Their interviews stressed high production values, and they had trained interviewers and professional videographers using broadcast-quality video, audio, and lighting equipment. Each interview was tracked with an advanced database that logged initial contact, language of the interview, people assigned to it, location and scheduling of the interview, release forms, and the shipment and receipt of the interview tape.

When tapes were received at the Shoah Foundation offices, they were logged, duplicated, and digitized in a room that could handle ten interviews simultaneously. Digitized interviews were then viewed by teams of subject-matter experts, who would index and catalog the content by the time code on the video. The result of this work was that a person could go to one of the Shoah Foundation's networked workstations and type in a keyword—for example, the name of a Nazi concentration camp—and see a list of interviews. You could then click on the interview and immediately view a segment where the narrator talks about the camp. To see on-demand video working on such a massive level in 1995 was impressive and inspiring. It was also very expensive, as the Shoah Foundation was using mainframe computers and a massive hard-disk storage system that was supported by robotics to organize and move storage drives based on which interviews were in highest demand. Furthermore, the reach of this early Shoah Foundation system was limited, as it only connected to a few workstations on a fast local area network.

The cost of the Shoah Foundation's project was far beyond what Densho could afford. Even with its most optimistic fundraising projections, the team realized it would not be able to raise enough money to replicate the technical infrastructure and operations that they had seen. However, what excited Oki and Ikeda, because of their recent experiences at Microsoft, was the understanding that emerging personal computer technology and the Internet would eventually become low-cost substitutes, and improvements to what the Shoah Foundation was doing. Ikeda already had the experience of leading a team at Microsoft to design a personal computer system to create, preserve, organize, and display text, images, audio, and video during the creation of reference CD-ROMs at Microsoft, including a multimedia encyclopedia and dictionary. For these systems to be useful at Microsoft, the multi-format content had to be sharable amongst dozens of people across a local area network, similar to what the Shoah Foundation was doing. Although personal computers weren't powerful enough to play high-resolution video, and the storage requirements of video were expensive, these systems would get faster and cheaper, as it was possible every two years to buy twice the computing power and storage capacity for half the cost. Furthermore, the World Wide Web was emerging as a powerful network for sending multimedia content easily and cheaply to the general public.

Although the passage of time was a friend when it came to making technology better and cheaper, it was an enemy when it came to doing interviews. Japanese Americans who had adult memories of the World War II camps were quickly disappearing. An early decision of Densho was to postpone the creation of the technical infrastructure and to focus on doing interviews, collecting diverse stories of what happened to Japanese Americans, but designing the interview program in a way that would take advantage of a multimedia, digital future. Ikeda volunteered his time to become Densho's founding executive director, and

Oki contributed $1 million for start-up costs to conduct interviews and create a working prototype.

Starting the Densho Interview Program

Densho spent most of 1996 researching, planning, and training for its interview program. The planning process took much longer than expected, as the group found that it needed to consult with a wide range of individuals including oral historians, archivists, librarians, historians, sociologists, psychologists, counselors, technologists, TV broadcasters, journalists, documentary filmmakers, videographers, intellectual property lawyers, nonprofit managers, and Japanese American community members. A discussion with one specialist often ended with the realization that the group needed to talk with experts in yet another field.

The initial scope of the interview program was to identify and recruit a diverse group of Japanese Americans who were incarcerated during World War II and who currently lived in the greater Seattle area. During the planning meetings, there was discussion that Densho should focus on the entire community and not limit the interviews to prominent Japanese Americans or just success stories. It was decided that the goal of Densho should be to create a collection with as many perspectives as possible and to seek out and recruit narrators who could tell everyday stories or "stories less told" and who might be reluctant to be interviewed. In particular, there should be an emphasis on getting the women's story of the Japanese American incarceration and the story of individuals who acted against government actions.

The content focus of the interview program was on the World War II incarceration. However, to examine the impact of the incarceration on individuals, Densho adopted a life-course trajectory model for its interviews, where it was common to spend a third or half of the interview talking about life before World War II. The focus of the interview would then turn to the events after December 7, 1941, and how lives and life-course trajectories changed during the war and after.

An important decision with cost and operational implications was whether to record using audio or video. Densho decided to video-record the interviews based on two factors—the emotional impact of seeing people share their stories, and the future scenario when on-demand video would be common and expected. This decision increased equipment costs, with the need to buy and maintain expensive video equipment (camera, lights, tripod, microphones, etc.). It also doubled the number of people conducting each interview and added a distracting presence by having a videographer with a big camera present during the session. Furthermore digital storage costs—at the time, Densho's highest equipment-category cost—were increased dramatically by storing audio/video files rather than the much smaller audio-only files.

Another decision with cost, processing, and user interface implications was whether or not to transcribe each interview. Densho considered the video recording as the final product of the interview, and some members of the group believed it would be better to use the organization's limited resources to do more interviews versus doing transcripts. After discussing this issue with experienced oral historians, Densho decided to transcribe all of its interviews because transcriptions gave narrators an opportunity to more easily review and correct the information in the interview, which would increase its accuracy and value. A transcript was also useful for Densho video users, as they could preview a transcript before deciding to view a video, read the transcript while viewing the video, or use the transcript as a searchable finding aid to find a topic within the video interview.

Because of the public nature of making the interviews available over the Internet, it was important to discuss privacy issues with narrators and make sure they were comfortable with the content in their oral histories. Prior to recording, Densho members would explain how the interviews could be viewed from anywhere in the world by anyone. After the session was over, narrators were given copies of their videos to review before being asked to sign a release form.

In late 1996 and through the fall of 1998, the group did its first set of 93 recordings with 82 different narrators, conducted by 21 different interviewers. The interviews ranged from an hour to 15 hours in length (the latter over 6 sessions). The narrators were a diverse group. Women made up 46 percent of the narrators, and the ages of narrators ranged from a person who was 42 years old when World War II started (born in 1899) to a child of a person who was incarcerated (born in 1955). There was also a wide range of experiences from this first set of interviews, including military veterans, draft resisters, politicians, community activists, stay-at-home mothers, and non-Japanese Americans who witnessed the removal.

Creating a Prototype for the Video Interview Collection

In the summer and fall of 1998, Densho designed and created a working digital prototype for the video-recorded oral history interviews. The group felt that with the first 93 interviews, they had collected enough to create an authentic challenge of finding and viewing the information from an assortment of individuals with diverse perspectives. Densho also wanted to demonstrate to, and get feedback from, the Densho narrators and the Japanese American community on how the collected interviews would be used.

For the prototype, the interviews needed to be individually indexed so that users would be able to find information by topic. A scholar in Japanese American history was hired to create a controlled vocabulary for the topical index that centered on the World War II Japanese American experience. The index was used to tag the subjects and events in the interviews, with each containing multiple tags. To help users quickly navigate to their desired topic, the interview was divided into a series of short video segments, generally between three and seven minutes, based on natural topic breaks during the interview. Each section was titled and tagged with the appropriate subject or subjects. This made it easy for a user interested in, say, reactions to the bombing of Pearl Harbor, to quickly generate a list of relevant interview clips. He could then watch the streaming video of the clip and read the transcript of the segment.

In the late 1990s there were a variety of image, audio, and video file formats from which to choose. Most of the formats were only a few years old, with no clear digital archival standards in place. Densho took a pragmatic approach of making multiple back-up clones of the original video files and storing at least one archival copy on a stable medium at an offsite location. After archival masters were preserved and backed-up, a mezzanine file (a high-resolution digital file from which working copies could be made) was created. These were generated when the video interview was divided into segments, that were, on average about five to six minutes in length. The mezzanine files were combined to create a video copy of the full interview for the narrator (generally a VHS copy in the 1990s, and later a DVD), as well as lower resolution video files for the prototype and web.

Over the years, Densho has changed the formats for both the mezzanine and display files. For the latter, Densho started with RealNetwork streaming formats and Windows Media formats. As viewers' expectations for video quality increased, Densho changed its web display video to a higher-resolution Adobe Flash format. For mezzanine files, Densho went from MPEG-1 files in the 1990s to larger, higher-resolution MPEG-2 files as storage costs decreased. This move also addressed the need to create higher-quality display files.

The prototyping process was iterative. Online pages were initially designed by a graphic designer and then critiqued. An example of a change that was made during this process was an initial design that contained so much information that a user would have to scroll down a page to look at the transcript but then would be unable to view the video at the same time. The design solution was to view the transcript in a pop-up window next to the video when a user wanted to see the transcript displayed. After several iterations, basic design features like navigation methods, content to be displayed, and placement of multimedia elements were finalized. A database containing the meta information for each interview and segment was used to populate each page of the prototype with information like narrator's name and biography, interviewer's name, date and location of interview, notes, and locations of transcript and interview files.

To gain usability information, the working prototype was used as a museum kiosk on the exhibit floor of the Wing Luke Museum of the Asian American Experience in Seattle in 1999. This was the first public display of the interviews. Densho installed a video camera (with appropriate notices warning those using the kiosk) to track how people used the prototype to provide usability information. For example, what people looked at, where they spent the most time, what they did not look at, etc.

Bringing the Oral History Collection to the Web

In 2000, Densho started the process of bringing the oral history collection to the web in a way that the online collection could grow and connect with other primary source materials. A decision was made to design the online archive of oral histories to include interviews from other groups in order to help diversify the geographic scope of available oral histories. Densho approached other organizations and filmmakers with Japanese American interview materials, asking them to make their collections available. Over a dozen groups and documentarians agreed to this arrangement. As of 2014, about half of the 800 interviews available from the Densho digital archive come from other collections. This percentage will increase as efforts expand to identify and digitize older interviews—primarily audio only and conducted in the 1970s and 1980s by other organizations.

Densho also decided to complement the video interviews with historic photographs, documents, and newspaper articles about the Japanese American experience that Densho had collected (Figure 7.1). Some of these materials were scanned from the private collections of the narrators, helping to illustrate the people and places described in an interview. Other materials came from institutional sources like the National Archives and Records Administration (NARA), the University of Washington Library's Special Collections, and the Wing Luke Museum of the Asian American Experience (Figures 7.2, 7.3, 7.4).

Figure 7.1 Screenshot from Densho visual history collection.

Figure 7.2 Screenshot showing other visual history collections in the Densho Archive.

Figure 7.3 Screenshot showing photograph from Dorothea Lange Collection from the photo/document collections in the Densho Archive.

Figure 7.4 Screenshot using topics, looking at food in a World War II concentration camp. The table at the bottom shows photos, documents, and video interview clips on this topic.

The topical index of Japanese American history created for oral histories was expanded with more subjects, which allowed it to be used with photographs and documents. This provided a simple way for users to browse the entire collection of interviews and historic photographs, documents, and newspapers by topic.

Use in Teaching and Research

When the website was launched it became a favorite with instructors who teach the Japanese American incarceration story, whether as part of Asian American history, US history, or other courses. The site had a large set of unique holdings, was free to use, was well indexed, and was always available, making it an attractive resource for teachers and students. Many educators recommended Densho to students interested in doing a paper or project on the Japanese American incarceration or other related topics, or they incorporated video clips from Densho's archive into their lectures.[2]

One historian who taught Asian American history classes at a public university in California assigned an oral history project to students as part of these courses. For those who lacked eligible narrators to interview, the instructor provided only one option: the life-history recordings in the Densho archive. Because the material was already prepared for them, complete with transcripts, the instructor required students who used Densho interviews to look at more than one person and to compare some aspect of their experiences. Overall, she was very pleased with the papers based on Densho's interviews.[3]

Another historian who taught an oral history course at a large public university used Densho's life histories to teach students about the oral history process. The instructor had students listen to a segment, then watch it, then read it to illustrate the different impressions each medium registers.

The Densho site has also been used as a resource to show how to use primary-source materials in the classroom. Over 600 teachers from 22 states attended Densho's teacher training workshops in 2012 and 2013. These courses examined the World War II incarceration of Japanese Americans through multiple perspectives, with the use of video clips from Japanese American oral histories, as well as World War II-era materials including a Dr. Seuss political cartoon, photographs from Dorothea Lange and Ansel Adams, a US government newsreel justifying the incarceration, newspaper editorials, and a confidential government intelligence report.

Densho's online interviews were also frequently used by historians and other scholars in the course of their research and writing. Among the many recent books that used and cited Densho interviews were Cherstin Lyon's *Prisons and Patriots: Japanese American Wartime Citizenship, Civil Disobedience, and Historical Memory* (Philadelphia: Temple University Press, 2011), Greg Robinson's *After Camp: Portraits in Midcentury Japanese American Life and Politics* (Berkeley: University of California Press, 2012), Gordon K. Hirabayashi, with James A. Hirabayashi and Lane Ryo Hirabayashi's *A Principled Stand: The Story of Hirabayashi v. United States* (Seattle: University of Washington Press, 2013), and Roger Daniels's *The Japanese American Cases* (Lawrence, KS: University of California Press, 2013). Numerous graduate theses also relied heavily on Densho interviews.[4]

Next Steps

The next evolution of the Densho online interviews is to encourage storytelling by its viewers. Densho wants users to share and explain interesting stories from the Densho visual history collection on social media platforms like Facebook. Densho also hopes that people will construct their own interpretations utilizing the online video collection—for example, students creating documentaries about the Japanese American incarceration or making linkages between Japanese immigration in the early 1900s to immigration today. To help make these things happen, Densho is redesigning its web technology to allow for social-media sharing and the downloading of editable video files under a Creative Commons License. In addition, Densho will create educational modules on the ethical editing of its materials. To help create historical context for the interviews, Densho is creating a 1,000-article online encyclopedia about the World War II Japanese American experience that will be completed in 2014. Interview segments will be tagged with encyclopedia articles to provide historical understanding and context when the clips are viewed or used.

Densho used oral histories to document the stories of the Japanese American experience in the twentieth century, centering on the World War II incarceration. Digital technology and the Internet allowed Densho to share these stories with millions of people from the comfort and convenience of their homes, schools, and offices. Moreover, through topical indexing and linking, users serendipitously found and browsed complementary stories, photographs, and documents deepening their understanding of Japanese American history and aligning with Densho's mission of keeping this story alive for future generations.

Densho is a model that other oral history programs are considering with the decreasing expenses of digital technology, the emergence of low-cost cloud storage, and the reach of social media. However, it is a model that requires an institution to embrace and react to the rapid change in technology, as new platforms emerge (World Wide Web, mobile devices, social media) and digital formats evolve. It is common for technical obsolescence to happen within a decade, so an organization needs personnel in place who are able to plan and prepare for the inevitable changes. The reward for taking this digital plunge is that your reach will increase as new and younger audiences find and engage with your materials.

Notes

1. In October 2014, Densho received the Stetson Kennedy Vox Populi ("Voice of the People") award from the Oral History Association. The award recognizes oral history-based work that creates a more just and humane world.
2. The author would like to thank Cherstin Lyon, Valerie Matsumoto, Lane Hirabayashi, Karen Inouye, and the many others who have informally shared with Densho their use of Densho materials in teaching.
3. Valerie Matsumoto, phone interview, October 3, 2013.

4. Among the many such theses are Robert Alan Hegwood's "Erasing the Space between Japanese and American: Progressivism, Nationalism, and Japanese American Resettlement in Portland, Oregon, 1945–1948" (MA thesis, Portland State University, 2011); Susan Elena Legere's "Narratives of Injustice: Measuring the Impact of Witness Testimony in the Classroom" (PhD diss., Boston College, 2012); and Caleb Watanabe's "Islands and Swamps: A Comparison of the Japanese American Internment Experience in Hawaii and Arkansas" (MA thesis, University of Arkansas, 2011). The last seems to be based almost entirely on an analysis of 25 interviews from Densho's archive.

CHAPTER 8

Deconstruction Without Destruction: Creating Metadata for Oral History in a Digital World

Elinor Mazé

Jacques Derrida had a hard time getting into college; he kept failing the entrance exams. In his last attempt at the exams, the budding deconstructor, given a page from Diderot's venerable *Encyclopédie* to critique, labored in his essay "to uncover a range of meanings fanning out from each sentence, each word." The examiners objected. One wrote, "Look, this text is quite simple; you've simply made it more complicated and laden with meaning by adding ideas of your own."[1]

Those of us whose job is crafting metadata for oral history collections cannot help but sympathize with Derrida. Even if we were not already disposed to the habits of postmodern deconstructionism, digital technology has brought us global clients, innumerable points of view, nearly limitless paths of access, and endlessly multiplying forms of representation. In the days when *information resources* meant books and journals on library shelves, the cataloger's job was simply "to enable people to locate items within the library's collection" by creating for each item catalog records that served "as facsimiles of the item itself."[2] Through one printed or rumored source or another, people arrived at the library either knowing what they wanted, or with a plan to browse shelves (or convince librarians or archivists to browse on their behalf), hoping to discover relevant new treasures. Now, widely shared metadata serves as advertising for the treasures, makes them discoverable to all and, as often as not, accessible to all. The metadata practitioner's job is to enrich our collections with pointers and descriptors of many sorts to bring interviews in all their formats to the attention of every user we can imagine, and to remain constantly vigilant for new points of view, new nuances of meaning and implication, and new uses of the stories we care for. We are professional deconstructors. But at the same time, we strive to remain committed to the principles of responsible archival practice, to standards of more than one academic discipline, and to the sensitivities of our storytellers and their communities, all of which call for meticulous preservation and communication of the original context in which the stories were told.

When metadata crafting began (and was generally known as *cataloging*, as to many it still is) and its purpose was to document what a particular library had on its shelves, a great deal of serious thought was devoted to how best to describe library items so that librarians and their patrons would know what exactly was housed there before making a trip to the shelves. Should they want to cast an eye over the shelves on a mission of discovery, the same system of description would serve to guide the arrangement of things so that similarly themed items were in proximity to one another. "What is it?" and "What is it about?" were both questions the earliest catalogers devised their systems to answer. They are still the questions to be answered today.

What Is It?

When the Baylor University Institute for Oral History first started to think about creating a catalog of its collection, it had already been at work recording interviews and transcribing the recordings for almost 15 years. The Institute's catalog addressed both of the questions identified above, but for the first, the "What is it?" part, the answer was implicit; it revealed more in a consideration of what was left out than what was included in each item entry. That each interview was, at its genesis, an audio recording made on a length of magnetic recording tape with a particular set of recording devices and technical settings was a fact given rather short shrift.

Fortunately, from the beginning, the Institute's policy was to preserve the original recorded tapes, to back them up with copies for use in transcribing and such, and to save both originals and copies. Storage boxes for open-reel tapes were labeled with basic metadata: name of interview participant, date of interview, and length of recording. Recording speed was usually noted, as was whether the recording was made in mono or stereo mode. Nowhere was there any record of such other facts as the brand and provenance of the tape or the machine on which the recording was made. There was no inkling at the time that this information would be important, providing clues, decades later, about the long-term survival of the medium and symptoms of deterioration it was likely to acquire, the sort of information that is now so crucial to future efforts to prioritize the many needy candidates for digitization.

Some of the information about each of the Institute's interviews had always been recorded. From the beginning, stage sheets tracked work. These were handwritten forms on which information about each interview was noted—names of interviewee and interviewer, date, place, and duration of interview—as well as the completed stages of transcript processing (transcribing, audit-checking, editing, binding, depositing in the archives). Physical description of the recording medium and technical details of the recording, on the other hand, were completely missing from these early records. This data existed only on the boxes and cases in which the tapes were housed, and, as noted above, it was sketchy.

It is not surprising, then, that in creating the first catalog, all attention was given to revealing the facts of the interview and the transcript. In reality, though, even answering, "What is it?" about the interview—as it was represented by its surrogate, the transcript—was a complex matter. The catalog used the term *memoir*. In the introduction, it was noted that "the terms *memoir* and *interview* are used interchangeably"[3] in the oral history world. It is not entirely clear that this was universally adopted practice even at the time, and in fact, the Institute's own application of the term was complex. The catalog's table of contents listed its primary sections as "Memoir Abstracts" and "Memoirs in Progress," and in the bound, deposited transcript volumes, the title page read, "Oral Memoirs of..." In practice (and to this day), it seems that *memoir* was used for the final product—that is, the completed transcript, reviewed by the interview participants and carefully edited, with front matter added, printed, and deposited in the archives. This final product might contain the transcript of a single interview or the collected transcripts of a whole series of interviews with a single individual. It was occasionally transcripts of oral histories with two or more individuals, recorded together, separately, or in varying combinations over the course of a series of interviews. Long volumes, those over 100 pages, usually included an index. The catalog listed memoirs, noting in each entry how many interviews were involved and their dates, the total duration of the interviews, the total pages comprising the volume, and an abstract of the content of the whole.

In creating the first catalog of its collection, the institute faculty and staff devoted a great deal more of their attention to answering the question "What is it about?"

What Is It About?

A decade or so ago, a group of oral historians at another university asked us to participate in their project to locate and describe collections nationwide that documented the history of women immigrants. I was newly arrived at our Institute and took on the task as a way to help myself get better acquainted with our own collection. I thought it would be simple. Although we had no project specifically dedicated to women immigrants, I supposed that a small set of easy-to-identify oral histories would fill the bill. We had transcripts of every interview we had collected, but at the time, a large portion of the completed and publicly available ones—those done before the mid-1980s—existed only as bound volumes on the shelves of the archives. We had more recent transcripts in electronic formats, including a very great number that had not yet been completed and were unavailable to the public, but lacking only a bit of final editing and formatting. Armed with our two-volume finding aid, which covered the years from our 1970 founding through 1996, and very primitive searching tools to comb through the electronic texts, I set to work, expecting that a few hours would get it done. However, the more deeply and broadly I considered the topic and all the aspects of it that a researcher might have an interest in, the more it became clear that we had considerable riches to offer.

Our geographic area in Central Texas, our state as a whole, and a number of our broader areas of research—religion and culture, family and community life, and the tantalizing catch-all, "special"—were all woven through with the personal stories of women of myriad ethnic backgrounds. There were interviews with recent immigrants, those who had come here as children, those born here and growing up in close-knit, often non-English-speaking communities to parents freshly arrived, and all of these women had some possibly significant bit to tell about the American woman's immigrant experience. It seemed important to list them all, but if the list was to be of any use to anyone, it was obviously also important that each interview be tagged, and its relevance highlighted.

I had been given a simple worksheet to list what we had to contribute. I filled it in with the main facts about our collection, but I appended a huge spreadsheet full of all the detailed information about each relevant part of each interview. I was in the audience when the project leaders presented their project at that year's Oral History Association (OHA) meeting, and I saw their finished product—the worksheet information, as I recall, gathered from their participants into a reformatted master list. They had found no way, of course, to integrate the avalanche of information I had provided, so the Baylor University Institute for Oral History, with its partner archives, The Texas Collection, were simply listed, with information about where we were and how to contact us for access to transcript volumes. I offered a few comments during the discussion after the presentation, arguing that although diligent researchers would no doubt benefit from the exhaustive search results I had produced, they would probably really want to do for themselves what I had done and comb through everything on their own, looking for those unexpected trails through previously unexplored territory. We just needed to make the content available and develop better tools to browse and search that content—tools that should be universally accessible. What was needed, I argued, was a nationally available, publicly sponsored template for documenting, abstracting, and indexing oral history collections—something simple to use, with an easy pathway by which to add it to some sort of readily available database (this could even be on the World Wide Web!). Someone asked for specific ideas on how this wondrous national oral history resource might be created and maintained, and I had to admit I had nothing to offer except my eagerness to help. There was no outpouring of shared eagerness from others in the room.

This early experience illustrates two of the challenges in taking advantage of the opportunities—and meeting the expectations—created by digital technology. First, this discussion spotlighted how much work we needed to do to describe the items in our collection in ways that would make them meaningful, discoverable, accessible, and useful to researchers everywhere. Secondly, the conversation should have made it equally clear how important it was to keep in mind the need for the compatibility of any descriptive systems we used for sharing information. We undertook the first work vigorously, in efforts that continue to the present, scanning our typescript volumes and making the full text all of our transcripts available online, whether they had been scanned, born-digital, or were still in final draft stages of editing. We migrated our finding aid, which was a digital version of those early catalogs, to create interview-level records in Baylor's online library catalog. We moved from the homegrown schedule of very general descriptors (e.g., *education, civil rights, politics, religion*) to Library of Congress subject headings. Later, we moved the entire collection to the university's installation of CONTENTdm for digital collection management, and in the process, further refined and enhanced the metadata, which described each interview and its provenance in great detail.

As to the second challenge, the global compatibility of our collection of metadata, we began struggling with that relatively recently. We are working on how best to describe our materials for the benefit of those alerted to, and interested in, our collection, but coming to it through different portals than ours? It is gratifying to see that the idea of creating a core of metadata for oral history is now, finally, beginning to get serious consideration. Recent meetings of the OHA have included sessions aimed at beginning the discussion on what should constitute this metadata core.[4] The aim is to create a recommended set of data about interviews and interview projects which can be made available to those who record oral histories, who oversee projects, and who curate them in all sorts and sizes of libraries, archives, and independent collections. The challenge is to develop the channels by which such a core could be distributed so that the standardized metadata harvesting tools now available, such as those of the Open Archives Initiative,[5] can be used to make the existence and content of oral history collections widely discoverable.

The globalization of context is really what much of the current work to expand the role of metadata for oral history amounts to. This might be seen, understandably, by many interviewees and communities as a destruction of context as they perceive and intend it. As Steven Cohen has recently put it, this "places the burden on the transcriptionist to do a better job and make more effective use of technology to convey meaning."[6] This burden is not only the transcriber's, but also that of the crafter of all the other metadata by which an interview will be presented to the world. But the transcribers, the indexers, the abstractors, the assigners of headings and tags, all have biases, points of view, membership by birth, upbringing, education, or conviction, in one culture and community or another. There simply is no global understanding of things. Even within a single cultural context, the discernment of significance by another person—someone creating metadata or otherwise analyzing or studying an interview—can be startling, and even disquieting, to the interviewee. These days we occasionally encounter the reality of this as our collection of transcripts comes so much more readily to the attention of the public through digital access.

Not long ago, an interviewee asked us quite passionately to remove her interview from our online collection. Now in midcareer, she had discovered the transcript of her interview online and in rereading it in her present frame of mind, feared that some comments she had made about a family member would be hurtful to that person as well as possibly damaging to her own career. The interview took place over a dozen years ago, and at the time, the interviewee had signed the release form, had reviewed the transcript, and had quite explicitly noted her wish to be able to use it in writings of her own. But in those years, our Institute was still depositing bound volumes in the archives downstairs, accessible only by appointment with the archivist, and the world would know about an interview only by a very brief, bare-bones record in our university library's online catalog. Our agreement forms, however, did give us copyright to the material and made it explicit that we would make it publicly available. What caused her to reconsider her donation of her story to us, apparently, was the enlarging of context, both geographically and temporally, in which her interview was presented. She had discovered her interview more or less inadvertently in a Google search and realized her family members and her employer could easily do the same.

The interviewer, in this particular case, worked with her to find a compromise solution that would avoid consigning the whole interview to the dustbin and quickly agreed simply to expurgate the troubling part of the transcript. That particular bit, the interviewer felt, was not important to the story the interviewee told, nor to the aim of the project of which the interview was a part. It struck me, however, in reading the offending passage, that it was actually quite an interesting comment, in part because it echoed very closely what another interviewee had said several years earlier, for a different project about someone completely unrelated (although in a similar field of religious endeavor). The pair of comments, separated as they were by time and immediate context, seemed to me to suggest a possibly fruitful line of inquiry into an aspect of the broader topic both of these interviewees were addressing—in both cases, the difficulties and challenges facing women who chose careers in the institutions of conservative, evangelical Christian religious groups. Using tools now available or in development (and discussed a bit further below) which facilitate very granular subject analysis of both individual interviews and whole projects or collections, I could have highlighted for everyone this connection, this commonality of view between unrelated (and, most likely, not mutually acquainted) interviewees. Doing so, however, would certainly have further discomfited the later interviewee. Perhaps it was a connection of interest only to me, but are not such epiphanies concerning possible larger meaning and unexpected connections the essence of work in history? Is this not really why we collect interviews in the first place, as well as all the other primary sources we so treasure, and admire the work of those who bring fresh perspectives and insights to those sources? Relevance is always relative. Any metadata which documents the what-is-this-about—the so-called *aboutness*—of an object is always generated from a point of view; just as the interviewee cannot tell a story from anything other than her own point of view. Creating metadata is a process of deconstruction; that process will always be indelibly marked by the person who does it, and it will never be static across contexts and time.

In the traditional view, formal indexing and subject analysis for descriptive cataloging are performed according to "unbiased, systematic, and universal rules," and "a permanent or near-permanent correspondence between an indexing term and the *aboutness* of an article [or other archived item]."[7] Such a view is epistemologically indefensible, of course (this essay began with Derrida), but it still underpins most of what we metadata practitioners do when we answer the what-is-it-about question in preparing oral history materials for access. Our Olympian judgments in this regard have long been problematical, whether we acknowledged the problems or not. The sexism, racism, homophobia, and other biases of subject cataloging authority schedules such as the Library of Congress Subject Headings have been well reported and documented.[8] "It might be better to think of our coders as proxies for the general user, rather than as social scientists producing consistent data for analysis,"[9] said Michael Frisch and Douglas Lambert, in discussing early work to develop their sophisticated digital tool for analyzing and tagging audio and video interview recordings (discussed further below). Critics of established schedules of subject terms, often—significantly—termed *authority files*, ask who this "general user" could possibly be and how she or he could possibly have a proxy. Robert Perks has cited 1990s work by Canadian archivists Jean-Pierre Wallot and Normand Fortier, in which they argued that "both archivists and their users are active agents in creating their own values and meanings, finally putting to rest the concept of archiving as a neutral and objective process."[10]

The question, then, that follows "What is it about?" is, "Who gets to say what it is about?"

One of the choices much touted today for diversifying and popularizing the answers to "What is it about?" is broadly termed *social tagging*. Social tagging is manifest in the forms of the hashtags used on currently popular Internet text and image exchange services such as Tumblr, Instagram, Flickr, Twitter, and Facebook. What settles from a social tagging maelstrom is sometimes termed a *folksonomy* (a term coined by Thomas Vander Wal, who has since reported a certain discomfort with many of the uses to which the word has been put to describe any public tagging system[11]), and folksonomies have been much discussed and frequently deployed by librarians seeking to enhance the relevancy and user-friendliness of their catalogs and research tools.[12] Many have pointed out the shortcomings of public tagging as a generally applicable tool for characterizing *aboutness*, and where social tagging has taken hold and attracted widespread participation, site administrators, collection managers, or content analysts have often attempted to tame the mass of diverse and often seemingly random offerings with such devices as word clouds, ranking lists of frequency, or other metrics.[13] Social tagging's appeal is its service as a way to give exposure to those in the statistical long tail, the many diverse minority voices. What analytical tools do in taming the mass, in a sense, is make subject analysis a kind of plebiscite, and the result can be viewed as nothing but the tyranny of a fickle (or prejudiced or poorly informed) plurality, the same tyranny against which the victims of bias continue ever to struggle.

Newer metadata tools for oral history provide hybrid systems for subject tagging, or indexing, of interviews. These systems generally provide mechanisms by which both informal, common-language, ad hoc subject characterization and controlled vocabularies can be applied to describe interviews at a very granular level. Their implementation is most likely in organizational settings where a mix of amateur and professional metadata practitioners, groups more or less stable in composition for at least intermediate periods of time and sharing common direction and purpose, work to prepare interviews for public searching and browsing. A number of these systems are prominent at present. One offered by Michael Frisch of The Randforce Associates[14] applies a faceted classification system, or what they term multidimensional indexing (MDI), to interviews,[15] while another, the Oral History Metadata Synchronizer (OHMS), is now in final development by Doug Boyd and his colleagues at the Louie B. Nunn Center for Oral History at the University of Kentucky Libraries.[16] These systems are inspiring and offer great promise for helping researchers with broadly diverse interests find and use oral history resources. The systems are exhaustive and powerful, but they can be intensive in their demands on time, personnel, expertise, and the wherewithal to meet those demands.

But to return to the anecdote related above about the discomfited interviewee who wished to suppress, years later, what she had originally shared without reservation in her interview, there are actually two issues which this anecdote highlights, and both are products of the digital revolution. The first issue is the obvious one, occasioned by the move from local access to global online, from a bound volume on the Baylor campus, or a digital version available only by request or via Baylor-networked computers, to full transcripts accessible online anywhere Internet connections are available. The interviewee certainly did not have global access in mind when she agreed to give us her interview as little as ten years ago. We have to be sensitive to this as we devise policies for making interviews, and all of their associated metadata, available online.

The second issue is more complicated. Regretful hindsight, and, as often as not, a wish to revise one's own past, take back or change what one has said (to say nothing of what one has done), is almost universal among us humans. No doubt, there have been a great many interviewees through the years who, upon glancing again through the transcript of an interview given long before, have winced at some bit of what they said. However, volumes on shelves, or of very limited availability, are unlikely to attract attention from any but the most determined seeker as the years go by. Few probably would know that the interview happened or that a transcript of it existed, sometimes even if they were doing a determined investigation into a relevant topic. But now that digital technology has brought these personal accounts within easy and frequent reach—easy to search for as well as to stumble upon—it may be that we will see more of this impulse to revisionism (or suppression) on the part of those who told the stories.

Of course, there is nothing new about contention over the right to edit, revise, or censor oral history interviews. Oral historians have long debated, for example, whether interview participants should be permitted to review transcripts of their interviews, and if they are, which of the changes they propose should be effected in the final version. However, the digital revolution, and the resulting global exposure of oral history materials, has thrown the question into a much wider context. There is much current concern about privacy, in view of unfolding revelations of the reach and power of data mining tools in the hands of commercial and government entities, for purposes by no means transparent or subject to the will or agreement of the individuals whose information is mined. This concern is very real for the curators of oral history and the narrators whose stories they care for. The metadata we create for oral history interviews, and strive to make readily accessible, is certainly as vulnerable as other information in this regard. Nevertheless, in one respect, the impulse to suppress, delete, or revise one's story can be equally troubling. Most of the well-publicized attention to this right to be forgotten, as it is termed, is focused on such things as Facebook content. But archivists are concerned about the vulnerability to such a supposed right of some of the materials they care for, were that right to be vigorously protected. Some European archivists, for example, have reacted cautiously to popular moves to restrict access to some kinds of information, or to make it easier for those who wish to do so to delete information that they do not want made available.[17] Revisionism can be, and certainly has been, a dangerous practice, in service to tyranny of many kinds. That individuals might, either of their own volition or under duress, be able to change their stories to serve repressive political ends, is a prospect that should give pause to historians as well as the archivists who care for the primary resources so indispensable to the best practice of their endeavor.

There is no easy answer to the dilemma created by these complex sets of concerns. The issues raised will be debated for some time to come. If we are not careful, we may provoke more of the sort of response we at the Institute for Oral History received from one of our interviewees a number of years ago, when we presented her the draft of her interview transcript to review: "I've changed my mind, I don't want to be included in History, so please take my transcript and send it back to me."

Opportunities and Challenges Ahead

For Baylor's Institute for Oral History, moving from bound transcript volumes in the archives downstairs—available only by appointment with the archivist, and discoverable solely through a two-volume printed finding aid—to instant online access to full text of all transcripts with exhaustive descriptive detail has had revolutionary consequences, even though the move actually took place in evolutionary stages. The move has changed not only how we do things and how much it costs to get these things done, in terms of personnel time and the expenditure of the fairly static quantity of resources that we, like most academic organizations, are given to spend. It has also changed how the stories we make available are understood. The first changes have been obvious; the second are often subtle, and we are realizing them only gradually and often only accidentally.

Our tasks now are not only to define, describe, and preserve our collection of oral history interviews while striving to make them accessible, but also to enrich that collection, to broaden and deepen the understanding of each of its stories, to reveal many facets of their significance to diverse populations, and to do all of this in ways that are somehow likely to survive rapid, never-ending changes to all of the tools of preservation, access, and analysis. Our challenges are to keep pace with the concomitant need for skilled personnel, for continual training and retraining, and for unflagging quality control and the careful, vigilant management that requires. All of this requires funding, in an era of static, if not shrinking, allocation of resources to humanities-related enterprises. Technology has been the means of saving time and money in many spheres, but for our kind of work, it has meant that our need for support grows ever greater.

Like many in our discipline, I am neither an inventor of technology nor a globally informed seer of the future. Most of us are reactors. Today's technology presents challenges and opportunities to which we respond—creatively, we hope, and with some vision of both future and present utility as well as doses of wisdom and caution that help us select what to adopt and what to reject, what to accept as-is and what needs to ripen before picking. We must often be the first to see possible ethical and legal consequences, as well as historiographical ones. We have wish lists and a few high hopes, and we try to communicate these coherently and effectively to those with technical know-how greater than ours. However, the pace of advancement, or at least of change, has shortened the wisdom-acquisition timeframe. We have to wise up sooner, eye the horizon more eagerly, if not more anxiously. And there is inevitably competition, a need, to be an early adopter of the new—to show the way. This means a competition for funds, the ongoing expense of buying the new, as well as budgeting for its replacement, sooner rather than later. We must face the reality that replacement of technology means replacement of the human parts—the training, the knowledge and experience, the management, and maybe even the people themselves—as well as the hardware and software. The problem we face is how to plan, with strategies and budgets, for such an uncertain future.

All of these challenges notwithstanding, oral history is an enterprise of inestimable value. In my view, oral history cannot but prosper in its usefulness and value as the ready accessibility to its primary media—the audio and video recordings—improves. The digital revolution offers great promise for such accessibility. It also offers great promise for conveying not only the voices of narrators, but also the crucial information about them and the context in which they speak, beyond borders, to a global audience. Our survival on this increasingly crowded planet must surely rely on our great understanding of one another across geographical and cultural divides, an understanding that calls on our capacity for empathy, however challenged by differences of language, belief, politics, and history. Each joyful, grieving, musing, angry, or analyzing voice relating a personal story can nurture the exercise of that empathy.

Notes

1. Benoît Peters, *Derrida: A Biography*, trans. Andrew Brown (Malden, MA: Polity Press, 2013), 54.
2. Matthew E. Gildea, *Cataloging Concepts: Descriptive Cataloging; Instructor's Manual*, Vol. 1 (Washington, DC: Cataloging Distribution Service, Library of Congress, 2002), 8.
3. *A Guide to the Collection 1970–85*, ed. Rebecca Sharpless Jimenez (Waco, TX: Baylor University, 1985), xvii.

4. See, for example, the roundtable, "Love Metadata? Let Your Geek Flag Fly!" presented by Natalie Milbrodt, Cyns Nelson, and Steve Schwinghamer at the 2012 meeting of the Oral History Association in Cleveland, Ohio, on October 13, 2012. Also, ongoing efforts by Nancy MacKay and others to develop the oral history metadata core, summarized in Nancy MacKay, "'Oral History Core': An Idea for a Metadata Scheme," *Oral History in the Digital Age* (Washington, DC: Institute of Museum and Library Services), http://ohda.matrix.msu.edu/2012/06/oral-history-core/.
5. Open Archives Initiative, http://www.openarchives.org/.
6. Steven Cohen, "Shifting Questions: New Paradigms for Oral History in a Digital World," in *The Oral History Review*, 40 (2013): 157, http://ohr.oxfordjournals.org/content/40/1/154.full.pdf+html?sid=671add90-cbdb-4c2f-bee9-16717c580fcf.
7. David Woolwine et al., "Folksonomies, Social Tagging and Scholarly Articles," *Canadian Journal of Information and Library Science*, 35(1) (2011): 77–92.
8. See, for example, Hope A. Olson, *The Power to Name: Locating the Limits of Subject Representation in Libraries* (Dordrecht: Kluwer Academic Publishers, 2002).
9. Michael Frisch and Douglas Lambert, "Case Study: Between the Raw and the Cooked in Oral History: Notes from the Kitchen," in *The Oxford Handbook of Oral History*, ed. Donald A. Ritchie (Oxford: Oxford University Press, 2011), 346–347.
10. Robert B. Perks, "Messiah with the Microphone? Oral Historians, Technology, and Sound Archives," in *Oxford Handbook of Oral History*, ed. Donald A. Ritchie (Oxford: Oxford University Press, 2011), 323.
11. Thomas Vander Wal, *Vanderwal.net*, "Folksonomy Definition and Wikipedia," http://www.vanderwal.net/random/entrysel.php?blog=1750.
12. See, for but two of many examples, Jessamyn West, "Subject Headings 2.0: Folksonomies and Tags," *Library Media Connection*, 25 (2007): 58–59; and Woolwine et al., "Folksonomies," 77–92.
13. See, for example, Frisch and Lambert, "Case Study: Between the Raw and the Cooked," 343–344.
14. The Randforce Associates, LLC: Oral History and Multi-Media Documentary, http://www.randforce.com/default.asp.
15. Douglas Lambert and Michael Frisch, "Digital Curation Through Information Cartography: A Commentary on Oral History in the Digital Age from a Content Management Point of View," *Oral History Review*, 40 (2013): 135–153.
16. "Louie B. Nunn Center for Oral History," University of Kentucky Libraries, http://libraries.uky.edu/libpage.php?lweb_id=11&llib_id=13<ab_id=1370.
17. Eric Pfanner, "Archivists in France Fight a Privacy Initiative," *The New York Times* (June 16, 2013).

CHAPTER 9

"We All Begin with a Story": Discovery and Discourse in the Digital Realm

Mary Larson

In the interview he did for this book, Gerald Zahavi reached beyond the bells and whistles, the flashy platforms and the Flash files, the tagging and the metadata, when he observed that:

> Good oral history is still good research, good questions, building good rapport, creating a dialogic relationship, deep listening skills. All of those are still fundamental to good oral histories. What technology has done, is it's given us the ability to collect it in a way that's easily preservable and easily disseminatable.[1]

The authors of chapters five through eight—Marjorie McLellan, Gerald Zahavi, Tom Ikeda, and Elinor Mazé—have talked about both how they address making good oral history and how they address *making meaning* and *making conversations* out of that oral history. For interviews and collections to be meaningful to researchers, to communities, and to participants, they have to be discoverable, and for the knowledge transmitted to make a difference, there has to be discourse. This essay spends some time parsing those concepts and discussing some of the underlying themes that are common to the contributions in this section of the book.

I first became aware of the importance of these paired issues of discovery and discourse during the years I worked with William Schneider on Project Jukebox, which is ironic considering that this undertaking is actually presented in an earlier section of this book because of its focus on orality. At the same time, we were striving to keep voice at the forefront of oral history presentation, though, we were also working feverishly on trying to make material discoverable. As Bill noted, one of the reasons behind Project Jukebox's inception was a fear that budget cuts would result in a severely diminished (or nonexistent) staff, so the driving idea was to provide access and discoverability in the absence of live reference assistance. Being able to find oral histories was as critical to the process as was presenting authentic voices.

Discovery

My work with Jukebox also made me keenly aware of the importance of gauging discoverability to various audiences, and I think that this can encompass not just the identification of resources, but also the discerning of meaning through the addition of context and through curation, as discussed earlier. Thinking of discovery in the more standard sense—that is, in terms of being able to locate and access particular information—there are two primary issues at play. Both serve important roles in the potential democratization of access to materials, which, as noted earlier, is a common goal of many digital humanists and oral historians.

The first issue, of course, is metadata, which at first glance might seem straightforward but is actually incredibly political in nature. A case in point: while working at the University of Alaska Fairbanks (UAF), I had the requisite knowledge to be able to catalog oral history interviews using standard Library of Congress (LC) subject headings, but those terms were not always meaningful in the context of what we were documenting in Alaska Native communities. Appropriate classifications simply were not always available, and even if they had been, the language of the academy and the language of the community were completely different, resulting in people being alienated from the life stories they had been generous enough to share. This issue is not just specific to Alaska, of course. In a 2011 article on indigenous knowledge organization, Deborah Lee cites a study by Kelly Webster and Ann Doyle that found in interactions with First Nations communities in both the United States and Canada that "the lack of appropriate description and classification were significant blocks to access for Aboriginal people."[2] In the interview I conducted with her for this publication, Elinor Mazé echoed some of the same sentiments regarding the utility or feasibility of LC subject headings in most settings outside of academic contexts:

> Among the things you and I have talked about just recently, for example, is Library of Congress subject headings. And then there's the opposite end, of

> folksonomies and public tagging and comments and such things, all of which are terrifically useful, each to serve its own purpose. Again, in a perfect world you'd do it all. You would have your professional librarians doing Library of Congress subject headings so that when things appeared in, say, your university catalog or your public county catalog or in WorldCat, it would be beautifully... and sensitively and intelligently tied to all other materials of the same subject. But as we know, that's a very specialized line of work, doing that, and it's also completely useless for an awful lot of people and an awful lot of subjects. Ordinary human beings simply don't think like that.[3]

For institutions that decide not to use LC subject headings, for whatever reason, there are the further issues of whether or not to develop controlled vocabularies to provide some sort of standardization, or to instead use more common terminologies, or even open materials up to crowdsourced tagging. At Densho, as Tom Ikeda noted, the project took the rare step of hiring a specialist in Japanese American history to create their list of potential metadata terms, and for subject-specific projects, that is probably more feasible than it might be for programs that cover a wider-ranging array of topics. While good controlled vocabularies and thesauri exist for technical, preservation, and administrative metadata, many institutions are still slowly working their way through subject headings, and that is an issue that many oral historians hope will find its own center within the next five to ten years. As we move forward in this area, we need to do so intentionally, keeping in mind that the practice of generating metadata entails sensitivity to a wide array of audiences and their needs, and it is not a simple, mechanical act without consequences.

The second issue at play in the democratization of discovery and access is structure. In the same article in which Lee discussed problems of classification, she also broached the issue of information structure and how that can be a very culturally constructed mode of understanding the world.[4] Again, this was something that was a common point of discussion while I was working with Project Jukebox. When we held local meetings to find out what people wanted their village's oral history computer program to look like, the answers differed from one place to another, but one thing that was fairly standard was that communities wanted to be able to access material through something other than a traditional index. That sort of entry point to a database was (and still is) often viewed in Native American circles as being a very Western, academic approach to information, as well as one that is very linear, and Elders generally had very different ideas on how they wanted to explore the oral histories.

When I helped with the work on the Chipp-Ikpikpuk/Meade River Project Jukebox, which focused on an area on Alaska's North Slope, people chose to access materials via Elders' photographs or by way of linked place names on a

Figure 9.1 Screenshot showing the main menu for the Chipp-Ikpikpuk and Meade Rivers Project Jukebox with various avenues of access.

Figure 9.2 Screenshot showing how Elders' photographs were used in the Chipp-Ikpikpuk and Meade Rivers project as a gateway to accessing oral histories.

Figure 9.3 Screenshot showing the use of maps as an entry point into the Chipp-Ikpikpuk and Meade Rivers oral histories.

map (Figures 9.1, 9.2, 9.3). In other locations, a decision was made to have audio stories tied to images, displayed photo-album style. In these cases, the index was almost always secondary for community insiders, but it may have been more important for those from outside, as I shall touch upon in the section on discourse. The point remains, though, that as oral historians try to create meaningful interfaces for discovery, it is incumbent that they be responsive to all of their constituencies.

Discovery is a two-edged sword, of course, and oral historians have seen within it both promise and threat. On the one hand, easy accessibility can break down barriers to the transmission of information. While it can be very simple, and even natural, for academics to visit university libraries or government archives, these same settings can be intimidating and actively off-putting to those who don't normally frequent those locations. Yet it is those very populations—the under-documented and under-represented communities that might see these repositories as threatening—that oral historians often collaborate with, and it is to those populations that we have a responsibility when making material available. As oral historians have moved to digital presentation and computers have become more readily accessible in less formal public spaces, there are more options for those who are uncomfortable in larger archival settings, who are homebound, or who simply don't have the money to travel to repositories. Online oral histories democratize access to information in a whole new way. At least, they do if the technology gap isn't in play.

At the same time, however, there are all manner of ethical concerns to be taken into account. Some of these were dealt with relatively early on in the process of placing oral histories online, while others have arisen more recently, as definitions of what is public and what is private have become blurred due to everything from social media to government surveillance. What does widespread accessibility really mean? How does this type of public presence impact narrators in the long run? And what is entailed with truly informed consent? Within this section on discovery and discourse, Tom Ikeda talked specifically about how this conversation played out at Densho. Because they knew from the beginning that all of their materials were destined for online access, they took specific care to ensure that people understood the implications of this dissemination. They had intentional conversations with their chroniclers about what Internet distribution of their interviews might mean, which is becoming an increasingly common practice with many programs across the United States as online publication can now be anticipated from the beginning of a project.

The ethical issues noted above are all ones that have been part of the oral historical conversation for almost 25 years now, and while the specific discussions on ethics might change and evolve as new technologies are developed, there continue to be questions. One of my personal concerns is based on what I have sensed is a decrease in the overall level of conversation on ethics, which is a discourse in its own right. When oral historians first started to discuss placing interviews online, the actual technical process for doing so was relatively complex. Software was often buggy, throughput speeds were dicey, hardware frequently wasn't up to the task, and there were no HTML editing programs, so all coding had to be done by hand. It was a time-consuming process, it was frustrating, and it swallowed resources like a rapidly growing adolescent, so the whole undertaking gave people pause. Because of the difficulties inherent in the system, though, most early digital practitioners carefully considered *all* aspects of their projects, including everything from the hardware and software to the ethics.

I worry that the technological ease with which we can produce oral history websites today makes some people less cautious. Programs are not investing the same level of time and resources, on a relative scale, and with so much personal information already available online, it does not seem as anomalous or dangerous now (at least to some) to be placing oral histories on the Internet. I think there is also a sense that the ethics conversation has already taken place and that everything has been resolved, but to my eyes, the questions have only evolved, not disappeared.[5]

Not all oral histories must be made publicly available online, of course. If oral history is to survive as a meaningful and scholarly endeavor, there still must be room in the profession for controversial or sensitive interviews that would never become part of the archival record if widespread dissemination were a part of the plan.[6] There are also heritage collections in repositories around the world that have unclear legal rights, or thoughtful archives that are concerned for other reasons about their ethical rights to put materials online, but most have still found ways to make their materials discoverable and available.

Doug Boyd makes the distinction between the *repository* approach, where full interviews are placed online, and the *exhibit* approach, where a more curated selection of excerpts and related materials are presented instead.[7] Many of the programs that have reservations about online availability will tend toward the latter, and there are a number of creative approaches that have been used. For example, when I worked at the University of Nevada Oral History Program (University of Nevada, Reno), we made a set of excerpts available through a project titled *Nevada Voices*. Working with local educators from the Teaching American History Project (TAHP), we pulled limited audio and textual selections specifically tied to state history subjects that they were required to teach, and we commissioned TAHP instructors to design companion lesson plans. There is a wide range of similar options, of course. We also had an extensive online index to all the program's oral histories as well as a collection catalog, and many programs have taken variations on that approach, with differing levels of granularity.

Discourse

In looking at the essays in this section and the ensuing interviews, discussions around discourse seem to fall into two categories. There are the internal conversations that take place during the *creation* of oral history projects and their associated online presentations, and then there is the external discourse that is engendered by the *reception* of a project once it has been made available online.

The internal variety involves discourse with participants in an oral history project—and that may include either the community where interviews are taking place or any collaborating stakeholders from outside of the community. The conversation with narrators and the group to which they belong needs to begin at a very early stage, so that oral historians can discover what information people (individually and in groups) want from a project. What are they interested in finding out about, and what resources would they like to see result from the project? This approach is something that I learned early in my career at UAF, and it is something that I have encouraged in my work, both at the University of Nevada,

Reno, and Oklahoma State University (OSU). If you want to have the trust of a community, you need to think beyond your own programmatic needs to be able to see what the project participants hope to get out of their involvement. Only when both sides are similarly invested do you really get the true collaboration and shared authority that are so often discussed in oral history.

Communication with communities has to go beyond just the project planning stage, though. There is a requirement for an ongoing engagement with the individuals or groups in question so that they have input into the actual presentation of the project, after the materials have been gathered. This extends beyond discoverability, as described above, to the contextualization that has been addressed so well in a number of the essays. How do people want to present themselves in their own words—and in images and documents and other formats—and what is meaningful to them in terms of access and design?

At the same time, there needs to be continued discussion with participating partners from outside the community. This might refer to academic advisors, as noted in Marjorie McLellan's essay, where she details the collaborating humanists who worked with her and her students on various projects. Equally possible is that these outside stakeholders could be on the boards of funding organizations, from collaborating university departments, or technical partners such as campus or institutional IT support.

In both her essay and her interview, McLellan mentions the successful relationships that she has had with various technical entities on her campus, but as others have noted earlier, there can be tensions between IT units and oral history programs, even when they are accustomed to working closely and cooperatively. Priorities tend to be different on both fronts, and that can cause problems when people are working toward divergent ends. Conversations between technical support people and oral historians are necessary from the very outset of projects to reduce conflicts and misunderstandings later. Everyone needs to understand what the proposed outcomes are going to be and how they will fit into any larger institutional infrastructure.

That institutional infrastructure itself can often pose the most difficulties. As Schneider and Gluck have shown, there can be tradeoffs entailed with IT support, and this can hearken back to the differing priorities noted a moment ago. IT departments are looking for content management systems that can be easily and efficiently maintained while oral historians may be trying to individualize platforms for each project, and that may not be sustainable in the long run. In her interview, Elinor Mazé details the challenges inherent in more homogeneous presentation forms:

> We have been frustrated with the finished products available for putting the collection online. CONTENTdm has got, for example, lots of power. It's a sophisticated system—perhaps too sophisticated for its own good—and our

> experience with it is that it's not easy for the average member of the public, whoever that is, and that's a pretty big average to use. It's not what we would hope. Now, that said, I think it can be made to be much better than it is, and we have seen organizations that have deployed it with considerable success, but that takes even more expertise and a bigger tool box of programming skills and experience, and so on, that really would stretch our capabilities beyond the limit. We can't do it all. That sort of thing we can't do.[8]

In the end, it's really about the compromises that need to be made in the service of discovery, and then the act of balancing all of that. In order to have a system that one can hope to have maintained by collaborating IT departments, programs have generally needed to sacrifice something in terms of design and customization. How far the balance is tipped in one direction or the other often depends on the financial or staffing resources available to oral historians.

In the presentation of projects, there is always the question of how to make these materials meaningful to external as well as internal audiences. Many times these groups can be two distinct constituencies who require very different things in order to be able to derive meaning from the oral histories. This is largely where the contextualization that has been discussed so thoroughly in previous chapters comes in to play, both for the local and nonlocal communities, outsiders and insiders, the general public and the academic world.

In trying to determine how best to make material available in useful ways, some projects, like Densho, started this internal discourse early. They released their prototypes to a limited audience first in order to conduct usability studies that would inform further development of the design and structure. While Densho's approach to gathering information was very scientific and quantitative, other projects in this volume have discussed less formal processes of eliciting feedback, either from community participants or from the more general public. At the Oklahoma Oral History Research Program at OSU, we combine our assessment modes. We gather anecdotal evidence from emails, phone calls, and social media, while at the same time taking into consideration the metrics derived from various analytical tools, such as Webalizer (for CONTENTdm) and Google Analytics (for more general website hits).

An external level of discourse can occur once a project has gotten past the prototype stage, has found its legs, and has been let loose into the world. As a result of the dissemination of information and meaning, there can be new public, scholarly, or pedagogical conversations. While in some cases oral historians are trying to start public discussions, in other instances they are looking at an academic conversation, either with other scholars or within the classroom. Marjorie McLellan, Gerald Zahavi, Tom Ikeda, and Elinor Mazé all noted the importance of their work with students in one form or another.

In McLellan's case, she was trying to engage a wide range of students, from middle school through graduate school, in a discourse about historiography and sources and what it means to do historical work, teaching them how to critically analyze sources and participate in the documentation of history. Ikeda, on the other hand, had a much more specific goal, and that was to start a conversation about civil liberties and to promote awareness of what could conceivably happen again if the general public is not reminded of our history. Hearkening back to the earlier discussion on key themes in digital humanities and oral history, the discourse, as Ikeda and the Densho board envisioned it, was about democracy. As he noted in his chapter:

> "What would happen if Japanese Americans who were unjustly incarcerated during World War II shared their stories?" This question shifted the discussion from simply conducting and preserving oral history interviews to a much broader discussion of how the stories could be used to promote and advance the country's democracy... In particular, it would be valuable for teachers and students as they were learning about democratic ideals and the Constitution to have access to these first-person stories of formerly incarcerated Japanese Americans.

Zahavi has been interested in changing the discourse with his students along technological lines. He has been trying to alter not just the conversation, but how it takes place, encouraging those in his public history and history and media classes to work not just in text, but also in images, video, audio, and compiled multimedia. In his interview, he related how he started to use other formats as a way of engaging students and getting them to think differently about their work:

> Since grad school, I was interested in how history is presented to broad audiences... I got into radio because of a technical interest and because I thought it was a political instrument. And then radio became very actively a part of public history when it was brought it in to the curriculum, when I brought it into the curriculum at the University of Albany. When I actually created a course to teach students how to use radio and how to produce radio documentaries or how to produce radio features focusing on history, then it was very consciously a public history mission that I had in that role at the university.[9]

He went on to note that the History and Media and Public History programs at his university have become so intrinsically connected that they are actually in the process of being merged. While that will take a year to implement, the change is already underway and may very well signal an important shift in the way public humanities are considered at academic institutions.

Before her recent retirement, Mazé worked with students in a very different context at Baylor University's Institute for Oral History, as she was trying to involve them in another kind of discourse. In her position as senior editor, she dealt with student employees on a regular basis, and, while teaching them about indexing and metadata, she attempted to get across to them a larger conception of the material they were tagging. Reflecting the earlier discussion on the potentially political nature of metadata, she wanted her students to be able to envision other potential audiences beyond themselves, so that they could understand who else might be using the material in the future and what these constituents might require for the interviews to be discoverable. She admits that it could be challenging:

> We've talked about how difficult it is to set relatively less experienced young people, people whose vocabulary is not very great, whose exposure to diverse populations isn't very great, whose reading is not very wide yet, to think beyond their own intellectual borders, if you will, to create key words or tags of some sort that would be meaningful to a wide variety of folks. That variety is so wide that you almost can't wrap your mind around it if the truth be known.[10]

At the same time that he was trying to engage his students in this discussion, Gerald Zahavi was also coming at this from another angle, trying to bring upper-level, scholarly, historical discourse into a new era by encouraging in his peers what he was promoting for his students—presenting history through a compilation of multimedia formats rather than simply through flat text. In his interview, he recalls the founding of the *Journal for MultiMedia History* and what that effort meant from an academic and historiographical perspective:

> Back in the 1990s, we formed a committee [at the University of Albany], a History and New Media committee, and I chaired that. There was a group of people in the department who sort of shared my vision of pushing history more into the realm of media and getting beyond academe and using the Web that was just emerging as a significant force... using it as a way of exploring both the expansion of history as well as exploring new modes of presenting history. The idea was to encourage some of the early work that was being done in multimedia publishing and hypermedia publishing and to put it within the framework of a peer-reviewed journal. We talked for a couple of years about this and finally decided to take off... The first issue was somewhat crude, but still gave you the idea. How you can actually talk about historical events... and actually hear people, the participants, hear their recollections, use oral history, and so be able to have an analysis as well as primary sources. Direct and intimate contact with primary sources in the form of oral histories, as well as visual—photographs and other elements in there. We take that for granted today, but it wasn't taken for granted then.[11]

As has been the case with many of the authors in this book, while Zahavi's was a unique and advanced approach, it may have been ahead of its time both intellectually and technologically. Only in the past few years have other journals, like the *Oral History Review*, been actively aspiring to these types of creative, digital compositions, and they are still running into the same difficulty that he reported in his interview—encouraging prospective contributors to think in terms of truly embedded multimedia.[12]

Earlier in this essay, I outlined three types of discourse to consider, and now it is time to look at the third, which is public discourse. It is clear that with all of these undertakings—the Miami Valley Cultural Heritage Project, the *Journal for MultiMedia History*, Densho, and the Baylor University Institute of Oral History—whatever their various perceived audiences were, one of them was the general public, and the interfaces have all been designed with that in mind. The authors have labored over making interviews discoverable through metadata and other approaches, and they have gone to pains to properly contextualize interviews and build online environments where the meanings of their oral histories stand the best chance of being engaged with and understood. They have also all taken their responsibilities to the communities they work with seriously, so that those groups are properly represented when the general public comes calling at their websites. By creating and caring for the projects they are involved with, they have made oral histories available to become part of the public discourse—to be viewed or read, discussed and critiqued, and to be enmeshed within the conversation of the world outside the academy.

As someone who very clearly identifies as a public historian, this sort of work was part of Gerald Zahavi's calling, and as he noted in his interview, he was on the lookout for ways to get scholarship out to the general population from very early on:

> At the same time, as recording technologies were evolving, a number of companies and technical consortiums had been developing media compression standards and formats that would allow for the delivery of media over the Internet... By the middle of the decade, it was possible to see a future in which both audio and video content could be integrated and easily delivered to users around the world. Excited about all of these prospects and deeply believing that all historians should be *public* historians, sharing their work with as broad an audience as possible, I began putting recordings on our servers and making them available to Internet users as streaming files, and soon as both streaming and non-streaming formats (MP3).[13]

This move toward public scholarship is, again, one of the characteristics that oral history shares with the digital humanities—this sense that information is for everyone and should be easily and freely available. Following this train of thought, one area where Zahavi was in the vanguard with the *JMMH* has carried forward rather significantly in the last few years, as the open-access movement has picked up an increasing head of steam in tandem with the technological changes that are making it more feasible. Back in 1998, this sort of approach was a much rarer bird, and while the multimedia format for the *JMMH* was unique at the time, I would argue that the peer-reviewed, open-access nature of the journal was equally revolutionary.

Today, librarians and other scholars around the world are leading the charge for unfettered access to publications, and two areas outside of libraries where the call has been loudest have been within the digital humanities and oral history communities. I believe the concept of free and open access is inextricably tied to the impulse for democratization that is shared so profoundly by the two groups. When paired with the commonly held trend toward public scholarship, those tenets are reflected in a number of ways. One peer-reviewed journal of particular note that became freely available to the public in 2008 is that of the Canadian Oral History Association, *Oral History Forum d'histoire orale*, whose site can be found at http://www.oralhistoryforum.ca,[14] and there have been discussions about opening up possibilities for other journals, as well. Within digital humanities, support for the open access movement is most unequivocally apparent in the *Digital Humanities Manifesto 2.0*, which addresses this issue quite clearly, stating, "**The digital is the realm of the : open source, open resources**. Anything that attempts to close this space should be recognized for what it is: the enemy."[15]

The concept of open access goes beyond journal publication, of course, and extends to usage and permission rights for oral histories, images, and other materials gathered for projects. One of the options currently getting the most traction is Creative Commons (CC) licensing, and Jack Dougherty and Candace Simpson have described the possibilities for this type of licensing in their *Oral History in the Digital Age* article, "Who Owns Oral History? A Creative Commons Solution."[16] The idea with CC is to allow chroniclers to retain their copyrights while at the same time allowing for their interviews to be made accessible to the general public. The Creative Common site (creativecommons.org) provides details on six different combinations of licensing, based on how people choose to deal with issues such as attribution, commercial use, and the terms for reuse or derivative works.

Among the projects represented in this book, Densho has taken a step in this direction, as they have already adopted CC permissions for the photographs and documents that their archive houses, at least whenever appropriate. In his interview, Tom Ikeda stated that they are also beginning to investigate the possibility of expanding access to their oral histories going forward, and however they choose to proceed, they are being intentional about looking at ways that they can responsibly make their materials more easily accessible for use in creative projects, particularly for students.[17]

The Means of Production...

While constantly evolving technology can be a gift to oral historians when wonderful new tools arrive on the horizon, it can also be a curse. One of the issues that is discussed throughout this volume is technical obsolescence, and it is certainly something that can hinder both discovery and discourse. In terms of discoverability, there are huge challenges posed by institutional moves to new content management systems or to different metadata schemata, and most oral historians have experienced the slings and arrows attendant to these upheavals, as Elinor Mazé discusses in her interview:

> [P]erhaps in common with a lot of other institutions, the ways in which we present our metadata have evolved and changed, and the systems we have used to do this and the systems through which we make our materials publicly accessible have changed and evolved. In many cases we were using new systems and learning how to use them and how best to deploy them and how most effectively to use all their features as we went along, so it was very much a seat-of-the-pants kind of thing—learn as you go. We changed our minds several times along the way about what the best way was to present our material, so that means in some cases going back and fixing old mistakes of judgment and in some cases just having what the catalogers used to call in the old days a *broken collection*. A broken catalog. And we'd say, "From this point forward we'll do it this way, where that point before it will be that way." Given world enough, and time, we'd go back and fix everything, but you don't have that.[18]

While some of these decisions may be made by programs because of pressure to conform to larger institutional infrastructures, many of these problems occur because of an intentional response that was meant to address changing standards in oral history practice. With almost all of our more crucial data now in digital formats and on digital platforms, merely trying to keep reasonably current with technology and standards can become a Catch-22, with rippling effects across systems. Change is inevitable, though, and many of the upgrades, conversions, and migrations that have given us headaches in the past have made our archives sounder and our preservation practices better in the long run. It is important, however, to be constantly vigilant and intentional about adopting new systems so as not to find oneself hitting a dead end after adopting a trendy but unsustainable app or piece of software.

A program's ability to engage people in discourse can also be seriously compromised if the presentation of oral histories does not conform to what are seen as the latest ideals for quality. In his essay, Tom Ikeda discussed how Densho changed its formats over the years in response to public expectations, while Gerald Zahavi reflected in his interview on the difficulties inherent in keeping up with file types and streaming technologies when attempting to maintain or repurpose older projects.[19] Public outreach is even further threatened, however, by a lack of institutional support, as Marjorie McLellan has noted. She had started the online Miami Valley Cultural Heritage Project (MVCHP) with her students while she was teaching at the Middletown regional campus of Miami University, but some time later, she received a position offer from Wright State University:

> I left for Wright State, but...I realized that I was the only one that was going to care enough to take care of this project [MVCHP], to make it grow...So leaving meant shuttering it, and it was so early in its development that I never got the chance to take it very far, so that was hard.[20]

As with many digital undertakings, without an active advocate on campus lobbying for its continued existence and upkeep, the MVCHP was allowed to pass into cyber-oblivion.[21] The lessons learned from that project, however, are carried forward in the continuing work that McLellan is doing with students at Wright State University and in her ongoing effort to use technology to engage both students and the public in historical discourse.

The final words in this essay come from Gerald Zahavi, with whom we started at the head of this chapter. In his interview, he made a very brief statement that I think may be one of the most concisely profound things I have heard in years. He said, "We all begin with a story."[22] The fact that we want to share our own stories, hear others' stories, share them, find them, interact with them, and talk about them is part of what makes us human. It is also what makes us oral historians, since that is primarily what we do in our line of work. Once people have shared their recollections with us, it is, in turn, our responsibility to make them available to others. When we find ways to make them discoverable in the larger digital realm, we make space for discourse to occur. At the end of the day, it all comes back to how we represent oral histories, their contents, and their meaning—whether through metadata, community discussions, scholarly engagement, public humanities, or pedagogical practice. We use technological tools to enable us to do what we need to do, but just as the rocks are beneath the water, there is meaning to be found behind the computer screen.

Notes

1. Gerald Zahavi, Interviewed by Doug Boyd, May 19, 2014, available online at http://www.digitaloralhistory.net.
2. Kelly Webster and Ann Doyle, "Don't Class Me in Antiquities! Giving Voice to Native American Materials," in *Radical Cataloging: Essays at the Front*, ed. K. R. Roberto (Jefferson, NC: McFarland and Co., 2008), 189, cited in Deborah Lee, "Indigenous Knowledge Organization: A Study of Concepts, Terminology,

Structure and (Mostly) Indigenous Voices," *Partnership: The Canadian Journal of Library and Information Practice and Research*, 6(1) (2011): 6, https://journal.lib.uoguelph.ca/index.php/perj/article/viewArticle/1427#.VB2gshb1bEw.

3. Elinor Mazé, Interviewed by Mary Larson, April 28, 2014, available online at http://www.digitaloralhistory.net.
4. Webster and Doyle, "Don't Class Me in Antiquities!"
5. For a longer discussion of this, see Mary Larson, "Steering Clear of the Rocks: Ethics and Oral History," *Oral History Review*, 40(1) (2013): 36–49.
6. A double session on this very issue was on the program for the International Oral History Association's annual congress in Barcelona in July 2014.
7. Doug Boyd, "Search, Explore, Connect: Disseminating Oral History in the Digital Age," in *Oral History in the Digital Age*, accessed June 28, 2014, http://ohda.matrix.msu.edu/2012/06/search-explore-connect/.
8. Elinor Mazé, Interviewed by Mary Larson, April 28, 2014, available online at http://www.digitaloralhistory.net.
9. Gerald Zahavi, Interviewed by Doug Boyd, May 19, 2014, available online at http://www.digitaloralhistory.net.
10. Elinor Mazé, Interviewed by Mary Larson, April 28, 2014, available online at http://www.digitaloralhistory.net.
11. Ibid.
12. For more on this, see the full version of Gerald Zahavi's interview with Doug Boyd on the website associated with this volume, http://www.digitaloralhistory.net.
13. Gerald Zahavi, Interviewed by Doug Boyd, May 19, 2014, available online at http://www.digitaloralhistory.net.
14. Janis Thiessen (editor of *Oral History Forum d'histoire orale*), email to author, May 29, 2014; see also Alexander Freund, "Oral History and Online Publishing: Establishing and Managing the Oral History Forum d'histoire orale," *Words and Silences: The Journal of the International Oral History Association*, 6(1) (2011): 12–17.
15. The Digital Humanities Manifesto 2.0, accessed May 3, 2014, http://manifesto.humanities.ucla.edu/2009/05/29/the-digital-humanities-manifesto-20/, emphasis and punctuation original.
16. Jack Dougherty and Candace Simpson, "Who Owns Oral History: A Creative Commons Solution," in *Oral History in the Digital Age*, accessed May 27, 2014, http://ohda.matrix.msu.edu/2012/06/a-creative-commons-solution/.
17. Tom Ikeda, Interviewed by Mary Larson, May 29, 2014, available online at http://www.digitaloralhistory.net.
18. Elinor Mazé, Interviewed by Mary Larson, April 28, 2014, available online at http://www.digitaloralhistory.net.
19. Gerald Zahavi, Interviewed by Doug Boyd, May 19, 2014, available online at http://www.digitaloralhistory.net.
20. Marjorie McLellan, Interviewed by Mary Larson, May 2, 2014, available online at http://www.digitaloralhistory.net.
21. Although as McLellan notes in her essay, the Miami Valley Cultural Heritage Project's existence has been documented and commemorated by previous publications, as well as by this one, now.
22. Gerald Zahavi, Interviewed by Doug Boyd, May 19, 2014, available online at http://www.digitaloralhistory.net.

PART III

Oral History and Digital Humanities Perspectives

CHAPTER 10

Swimming in the Exaflood: Oral History as Information in the Digital Age

Stephen M. Sloan

> Our inventions are wont to be pretty toys, which distract our attention from serious things. They are but improved means to an unimproved end...We are in great haste to construct a magnetic telegraph from Maine to Texas; but Maine and Texas, it may be, have nothing important to communicate.[1]
>
> —Henry David Thoreau, *Walden*

As this volume well demonstrates, the impact of digital media on oral history is wide and far-reaching. In a relatively short time, new technologies have revolutionized countless aspects of the work of oral history—from creation, to preservation, to use—and raised a multitude of discussions among oral historians on the impact of new technologies on oral history practice. What is also needed is a discourse on the nature of oral history in the midst of this dramatic change. In the revolution brought about by the introduction and rapid evolution of the digital age, what is the place of oral history as information in that new environment? As well-known professor and management consultant Peter Drucker declared in 1999, the first phase of the IT (Information Technology) revolution was focused on the "T" rather than the "I." In the new millennium, Drucker argued, the most pressing question that must be dealt with is the nature of information itself: "What is the MEANING of information and its PURPOSE?"[2] For this conversation, I would argue that oral historians need to follow the same path. It has been important to examine the technological aspect of this revolution, but what about the meaning and purpose of oral history as part of this new information landscape?

Contemplating the qualities of oral history as knowledge and information is an analytical exercise not unfamiliar to oral historians. For decades, much discussion centered on comparing the merits of data gathered through oral history to more traditional source material, particularly within certain academic disciplines. Since the professionalization of oral history in the mid-twentieth century, the strengths and weaknesses of oral history versus other research approaches have been a subject of debate. It is still an ongoing discourse.[3]

What is required now, however, is a broadened field for the consideration of oral history as information or knowledge. The scope of the discussion of the merits of oral history as evidence needs expanding, especially with the dramatic metamorphosis of information since oral historians first framed this discussion in the 1960s and 1970s. We need to contemplate questions such as "What is the worth of oral history in this age? What does oral history mean in an environment where information is radically more accessible and available?" This necessitates thinking through the nature of oral history as knowledge within this new setting. To frame this discussion, it is useful to have a more complete understanding of the contemporary information environment.

Digital Information Landscape

In many respects, our society went from analog to digital at the speed of light. Just over a decade ago, only a quarter of the information stored globally was digital. A significant majority of recorded information was held in analog formats, including paper, film, tape, and vinyl. Since that time, however, the proportion of analog to digital quickly reversed. By 2013, the explosion of digitally stored information meant that now less than 2 percent of global data is non-digital. In the span of just a few decades, all members of the G7 group—Canada, France, Germany, Italy, Japan, the United Kingdom, and the United States—were transformed into information societies where now, at least 70 percent of gross domestic product relies on intangible goods (information related) rather than material goods (physical agricultural or manufacturing production). As information philosopher Luciano Floridi notes, these are societies whose "function and growth requires and generates immense amounts of data, more data than humanity has ever seen in its entire history."[4]

This rapid eclipse of analog by digital can only be understood through recognizing the character of digital data in the new millennium. Our time is increasingly characterized by a torrent of digital data, whereas information in the not-so-distant analog age grew quite slowly. In 2003, researchers at Berkeley's School of Information and Systems posited that humanity, from the beginning of recorded history to the popularization of computers, had accumulated approximately 12 exabytes of data.[5] With the flood of digital data produced in just the few decades since the advent of the computer age, that total is now approximately 1,200 exabytes (or 1.2 zettabytes), the approximate equivalent of 8.18 million new libraries the size of the Library of Congress. We stand on the precipice of replacing the buzzword "exaflood" with "zettaflood" as a neologism for what Floridi calls the "tsunami of bytes that is submerging the world."[6]

Several forces have contributed to this deluge from all quarters. In December 2004, Google Inc. announced its ambitious Google Print Library Project, now Google Books Library Project, a large-scale partnership with several institutions to digitize and make available millions of library titles. By 2013, the project had digitized more than 30 million books, an impressive number, especially considering that total represents approximately one-fourth of the books ever published.[7] Google's book digitization initiative reflects the broad frenzy over the last ten years to convert everything analog to digital.

As the mind reels over the scale and scope of digital data mass, it can easily fail to notice the ways in which this trend operates at the individual level. Each of us has a rapidly growing digital footprint, which we create both actively and passively. Actively, the dramatic increase of participation in digital culture means more and more individuals are contributing digital data to the pool—blogs, comments, emails, social media, posts, etc. With the rise of Web 2.0 at around the midpoint of the last decade, the untapped potential for users to become creators of digital information became realized. Commentators were amazed at what a force crowdsourcing proved to be. One prominent example, among many, of what the collective gaze of individuals could accomplish was the online resource Wikipedia, billed as "the free encyclopedia that anyone can edit." In just seven years from the site's creation in 2001, Wikipedians pored more than 100 million man-hours into the creation and development of the platform.[8] As masses moved from passive viewers and users of digital information to creators, the amount of digital data dramatically increased, now doubling about every three years.

Taken together, digital authors provide 10 million images to Facebook every hour, post an hour of video per second on YouTube, and send 400 million tweets daily. The introduction of Facebook Timeline in September 2011 exemplified the new and expanding range of possibilities for information creation. This shift, though subtle, was significant. The default profile transformed from a list of most recent updates to a chronicle of an entire life since birth, a place, Facebook states, "to tell your story from beginning, to middle, to now,"[9] promising the definitive vehicle for historical self-representation, reaching back long before Facebook even existed.

Although many individuals actively build their digital footprint, it also grows passively, even if one refrains from posting, tweeting, commenting, or sharing. Through many avenues, a rising amount of information is harvested from individuals who participate and explore online. Web interactions are not anonymous, as data has become a key aspect of targeting user interest and purchasing in the intensely competitive virtual marketplace. As law professors John Palfrey and Urs Gasser point out, "In previous eras, third parties held information about individuals, but nowhere near the amount held in the digital era . . . the scope of what is being collected and the range of parties who are collecting it are both increasing."[10]

Despite these radical changes realized on the information landscape over the past couple of decades, society's seismic shift has only just begun. These are foundational transformations, but even in the most technologically advanced societies, as Palfrey and Gasser note, "no generation has yet lived cradle to grave in the digital era."[11]

Quantitative Conquest?

For researchers seeking proof, the developments described above present a pool of data that was once inconceivable. Investigators have eagerly embraced the opportunity to work with such bounteous data for statistical, mathematical, and numerical analysis. For those who have long argued the strength of a quantitative approach over qualitative methods, big data seems to be the final word in the debate, raising the question: If enough quantitative information can be gathered, is qualitative understanding still as important, or even, still relevant?

Advocates argue that possessing big data, as it is called, provides a new scope for researchers to understand the world in a way that is incomprehensible on a smaller scale. Big data has the capacity to tell us new things about reality that we were unable to see from the narrower, analog point of view. As information experts Viktor Mayer-Schönberger and Kenneth Cukier note, "The era of big data challenges the way we live and interact with the world. Most strikingly, society will need to shed some of its obsession for causality in exchange for simple correlations: not knowing *why* but only *what*. This overturns centuries of established practices and challenges our most basic understanding of how to make decisions and comprehend reality."[12] In other words, if we can more definitely understand "if *this* then *that*," the motivations/rationales/complexities underpinning those phenomena become less of a concern, or cease to be a consideration at all.

Within this idea, the way to a more definitive answer to the question "what is truth" is to muster more quantitative data around it. By way of just one example, consider an ongoing experiment researchers have undertaken with the mass of text digitized by the Google Book Library Project. These investigators ambitiously coined a new term for the approach: Culturomics. Culturomics is a field of study that uses the mass of information in the digitized books to track changes in language, culture, and history. Using a new digital lens, the

Ngram viewer, researchers can investigate trends in the written word en masse. Queries as diverse as comparing "slavery is" versus "slavery was," "Plato" versus "Aristotle," or "zombie" versus "werewolf" can be accomplished through searching for millions (or billions) of occurrences in the agglomeration of books and plotting those instances over time.[13] As Eric Aiden and Jean-Baptist Michel, the two researchers who framed the approach, contend, "new scopes are popping up every day, exposing once-hidden aspects of history, geography, epidemiology, sociology, linguistics, anthropology, and even biology and physics."[14] Aiden and Michel argue that the lens will "change the humanities, transform the social sciences and renegotiate the relationship between the world of commerce and the ivory tower," while the magazine *Mother Jones* countered by calling it, "possibly the greatest time-waster in the history of the Internet."[15]

Whatever the ultimate significance of Culturomics, the massive digital library created by Google and considered in this way demonstrates some key ways that the digital explosion of information has reinforced the drive for more quantitative analysis. Books, as they have been traditionally known, are very linear and quite statically framed within Culturomics. The information or understanding that they provide are firmly locked within a context that extends from cover to cover, or possibly from volume to volume. Although the content may remain the same in a digital iteration, digitized books are fundamentally different from their analog predecessors. The form changes their basic nature. As digital objects, they are now searchable, quickly excerpted, decontextualized, and easily tasked to another purpose. That is, they are easily manipulated. Digitized, they now give the user significant control over the meaning and purpose applied to the content of the digital object, whereas, in the analog age, the author held the majority of control.

As in the example of books, recognizing the implications of the transition of oral history from an analog to a digital object is vital. This is especially true when viewed within the radically transformed digital information landscape described previously. For oral historians, understanding the meaning of their work within the pool of digital information allows for a better understanding of why we do what we do.

The Immutable Core

Although oral history in digital form is a markedly different object than it is in analog, distinctive qualities essential to oral history persist into the new age. Many of these characteristics exist in tension with the overall character of the digital information pool. Exploring points where these tensions lie, especially related to features at oral history's immutable core, reveals what is distinctive and essential about the methodology for the new age. These are the attributes that distinguish oral history from the exaflood of other information available digitally. As we create, preserve, and share oral history, it is important that we hold fast to the form and function at the center of oral history. These qualities must not be surrendered in the ardor to realize the enhancements digital tools can bring to oral history work.

One essential element of oral history is that it is a linear, long-form source. As stated in the *Principles and Best Practices of the Oral History Association*, the average length of an oral history interview is two hours, and interviewers need to work to ensure there is "sufficient time allowed for the narrators to give their story the fullness they desire."[16] The long form of oral history allows for gathering extensive experience and documenting fuller understanding. Even in interviews that are framed fairly narrowly in a thematic sense, oral historians gather context, and the order in which events are related is important. Placing high value on the narrator's telling of a broader life history, oral historians work to document the frame of reference or circumstances that help shape the interviewee's perception and understanding of events.

As a long-form source, oral history's design is challenged by the manner in which digital objects are easily manipulated. There is a push and a pull at work here. It is a push driven by the capacity of new applications to easily segment and creatively present pieces of interviews, whether the form is a transcript, audio, or video. The pull to break up the long form comes from popular trends in digital information. Digital users increasingly consume and seek information in shorter and shorter increments. Reflecting on this trend in his essay "How Long Will People Read History Books?," historian William Cronon noted, "Twitter—with its 140-character 'tweets'—currently stands as the extreme symbol of this movement toward brevity."[17]

In an age where digital information is commonly characterized by its truncated delivery, the long form of oral history is increasingly important. As William Schneider has already noted, context matters, and, as oral historians know, it can mean everything. By definition, gaining understanding takes a measure of patience that entertainment does not. Not all users are willing to commit the time. The quick and nonlinear approaches to moving through information have made users, in general, less willing to be patient. One recent study of online video viewers revealed that users begin to abandon a video if it takes more than two seconds to start up.[18]

The long form of oral history matters. Oral historians must resist succumbing to the pull to brevity in sharing interviews online to the point that this impulse distorts the integrity of the interview. Although the mutability of oral history has increased with the turn to digital, we, as oral historians, need to work to resist bending our aims to popular culture's desires. Understanding requires a level of patience and deep listening from users that entertainment does not.

As oral history is segmented and excerpted, it moves further away from the co-authored piece built by investigator and narrator. There is a tension between preserving the meaning the interview held for the co-authors and remaking it in a form that suits the preference of the user. In creating oral history, we are intentional about allowing the narrators to tell their own stories. We must match that commitment in the way we disseminate that information to others as a digital object. When we do present segments or excerpts of oral history, we must think creatively and intentionally about the ways in which we preserve the context of the selections.

It is a challenging task in an age where patterns work against it. Surrounded by the mass of information available digitally, users are becoming habitually accustomed to approaching information in a nonlinear fashion. Journalist Nicholas Carr's book, *The Shallows: What the Internet Is Doing to Our Brains*, created a stir when it raised the question of how online activity is altering our mental habits. In reflecting on his relationship to information that he now accesses and moves through digitally, Carr remarked, "Once I was a scuba diver in the sea of words. Now I zip along the surface like a guy on a Jet Ski."[19] Oral history offers the opportunity for users to move vertically, down into experience to gain understanding, while the popular trend in Internet use is to skip vertically from item to item, a task for which the hyperlink was expressly designed.

Another essential element that rests at the immutable core of oral history is the curatorial role of the oral historian. As digital information has grown, availability and access have become less of a challenge and these concerns have been replaced by the larger task of navigability. How does one make sense of the exaflood? As digital humanities scholar Stephen Ramsay states, "That much information probably exceeds our ability to create reliable guides to it."[20]

Oral history, by its nature, is a source with an informed and purposed guide. Interviewers are accustomed to positioning themselves between research question and narrator, or inquiry and subject—a task that calls for a more reliable guide than we currently possess in the digital age. The deep significance of this curatorial role becomes clearer when we begin to understand the other forces shaping and selecting what information is provided or made available to our queries of the pool of digital information.

When a user is online and asks, "What is worth knowing about X?" there are many forces influencing the response that the question may receive. We have incorrectly made the word "Google" a verb synonymous with asking in an open search for information, which is increasingly no longer the case. It is an inquiry that is actually quite directed. Three elements are privileged in determining the information you receive when you ask online—what is profitable, what is popular, and what it presumes you would most likely want to see. All of these have implications as we think about the distinctive function of oral history as an element of digital data.

Although we often lose sight of its repercussions, a primary driving force behind the structure of online activity is financial in nature. This reality impacts the broad range of providers of digital information, not just for-profit concerns. In online terms, hits are the new currency. Often this can translate to real lucre, but it always means visibility, importance, and relevance to engender financial support and development. Designers and operators work to create interest in their website, to raise their profile and drive traffic to their domain.

In the current climate, competition for user attention is intense, and vendors have increasingly sophisticated tools at their disposal to direct consumption. Standing at the gateway to information is an especially powerful position. Even Google, a default brand that we have come to trust so much as our guide, manipulates the way in which we explore the digital world. The place we have conceded to Google should concern us more than it does. As media scholar Siva Vaidhyanathan writes, "Where once Google specialized in delivering information to satiate curiosity, now it does so to facilitate consumption."[21]

Just as financial considerations mold the search patterns of the Internet, another primary shaper of virtual discovery is what is currently popular. In general, when you search online, information that is currently trending, or popular, will be privileged in the answers given to your query. One can think of the interaction not as viewing a static set of digital information, but glimpsing a quickly morphing mass—following what is trending and generating hits, constantly reshaping in response to fluctuating popular interests. And, of course, by privileging popular items, Internet searches then in turn make those items even more popular. This helps explain how items can go viral so quickly.

The forces of trending and the shrinking of the news cycle wrought by the digital age have led to the prominence of what journalist Bill Wasik calls the nanostory. A nanostory is a story that goes viral and then disappears almost as quickly as it came. For a moment, Wasik states, "we allow ourselves to believe that a narrative is larger than itself, that it holds some portent for the long-term future; but soon enough we come to our senses, and the story, which cannot bear the weight of what we have heaped upon it, dies almost as suddenly as it was born." Wasik states that popular attention imbues the nanostory with something more significant than 15 minutes of fame; it grants 15 minutes of *meaning*.[22]

In trying to discern the characteristics of digital information that make items become popular and take on a life of their own, repeated research shows that those that shock or inspire are more likely to be shared and have greater capacity to gain viral traction.[23] Traffic is drawn to hyperbole and items that are sensational, brief, and vogue, and those objects are redistributed at a lightning pace.

A final trend to be aware of for our purposes is the ways in which our individual relationship to the data mass has taken a dramatic turn in recent years and become much more intensely personalized. The Web is no longer an anonymous and neutral pool of information waiting to be explored. The more information that we provide through a range of interactions with the virtual world, the more digital providers learn about us, increasingly collecting and analyzing our personal data. Knowing more about you increases their ability to cater to your desires more efficiently. As Internet activist Eli Pariser points out, "More and more, your computer monitor is a kind of one-way mirror, reflecting your own interests while algorithmic observers watch what you click."[24] It is a new phase of our relationship to the digital world. The cutting edge of this phenomenon is technology's ability to know what you want from it before you ask, whether it be an article, website, or product. The more data mined from you, the greater the accuracy in its matrix of presumption. In December 2013, Amazon secured a patent for "anticipatory shipping," a system that will allow the übershipper to initiate delivery of a package to a consumer before the product is even purchased online, sending what you want to buy even before you know you want to buy it.[25]

All three of these forces discussed above shape our discovery of digital information and are important when thinking about the characteristics or meaning of oral history as an element of the digital landscape. At their best, oral historians work for long-term relevance and significance in the information they collect, not an ephemeral perspective. There is a certain weighing of events/issues/ideas to judge what is of importance. The concern is not simply discerning the questions and answers being sought today, but determining what enduring questions we will be seeking answers for 10 or 50 years from now, an elusive but constructive pursuit.

The escalating personalization of online discovery is a subtle but significant change. While oral history is a process of discovery, an endeavor to understand difference so that we may pass on that understanding to others, this is not the case with the current trends in online searching. In other words, when we explore the Internet now, we are increasingly interacting with products, themes, issues, ideas, and opinions in which we, or those like us, have already expressed interest. Internet activist Eli Pariser calls this the "filter bubble." The filter bubble is "not designed to introduce us to new cultures. As a result, living inside it, we may miss some of the mental flexibility and openness that contact with difference creates."[26] As oral historians, we make fundamentally different choices in the information we pursue and the way we pursue it, what we privilege in the discovery process, and what we work to share. Masses of information do not necessarily impart new knowledge, and by no means do they convey meaningful and expanding understanding on their own, yet that is what we strive for when we present oral histories online.

A final and essential aspect of oral history's immutable core is the manner in which it provides rich and necessary qualitative understanding. In an age where the data pool continues to grow exponentially, one might think that there is a diminishing return in oral history's currency. On the contrary, the approach has increasing value in the information landscape. With the push toward quantitative analysis described earlier, qualitative understanding is needed more than ever. Even researchers who argue the vast potential of research through big data, such as Mayer-Schönberger and Cukier, warn of the dangers of a "dictatorship of data." As they explain, big data holds the threat that, "we will let ourselves be mindlessly bound by the output of our analyses even when we have reasonable grounds for suspecting something is amiss. Or that we will become obsessed with collecting facts and figures for data's sake. Or that we will attribute a degree of truth to the data which it does not deserve."[27] The pursuit of truth would be crippled without qualitative knowledge. As big data may increasingly shape the narrative, qualitative analysis brings the vital counter-narrative; a task oral historians have always been passionate about. Ironically, the virtual world, which seems to offer the ideal environment for counter-narratives to flourish, can also serve as an instrument for the perpetuation and hegemony of a dominant narrative. Qualitative analysis offers the opportunity for each voice to be heard, to encounter the exception, the nuance of existence, and the texture of reality. It is the deterrent to overgeneralization.

Conclusion

All would agree that new technologies have, of course, provided us with the ability to record volumes upon volumes of information about the experiences of individuals. It is a capacity that some have fully employed. One of the most intriguing characters the reader meets in Dave Egger's *The Circle*, a novel about the culture of Silicon Valley, is Stewart, who appears about halfway through the book. Stewart sports a wearable camera, one he has worn constantly for the past five years, capturing and cataloging a full record of his life. He is revered in the book as a courageous and inspirational figure for opening up his life to collective knowledge, endeavoring to be the "transparent man."[28]

Although a fictional character, Stewart represents a real trend in the use of wearable technology to document one's experience 24/7: lifelogging. Lifeloggers use recorders to capture everything they do in real time, creating terabytes upon terabytes of video record. The technology for the task is less obtrusive and more accessible than it has ever been. If a lifelogger misplaces her keys, she can rewind the tape. If she forgets a conversation, she can relive it. Why rely on memory when you can hit replay? On the surface, it may seem that what lifeloggers do is akin to oral history, but, if one looks closer, it becomes clear that the practice of lifelogging is much more about technology than it is about revealing the meaning of the human experience.

Much more prevalent than lifelogging is the phenomenon of self-presentation of life experience through social media. The rise of Web 2.0 has given each of us the ability to establish something that only a small minority had in the analog age, a public profile. All now have a vehicle available to them to frame and present a public persona that others can contribute to, like, or comment on.

Whether it be lifelogging, social media, or the other countless manifestations of the digital revolution at the individual level, oral history offers an insight into the individual that is increasingly distinct. Oral history is not merely experience recorded (as in lifelogging) or experience presented (as in social media), but it is experience *examined*—not merely documented and shared, but investigated with purpose, with the oral historian as an active agent in the process. We are at our best when we seek to reveal "why" and "how," which, despite their wonders, most digital sources of information fall short in providing.

Personally, one of my favorite moments when I conduct an oral history is when the narrator realizes something new about their own experience; connecting things heretofore unconnected, seeing patterns, context, connections, and begins grasping consistencies and inconsistencies in his or her own story. As oral historians, we do this on an individual level, but more broadly, we do this for the collective. It is a contribution more worthwhile now than it has ever been.

Notes

1. Henry D. Thoreau, *Walden* (New York: Thomas Y. Crowell & Company, 1910), 67.
2. Peter F. Drucker, *Management Challenges for the 21st Century* (New York: Harper Collins Publishers, 1999), 97.
3. In 2014, the Oral History Association will publish a formal statement on the merits of oral history as scholarship to be used by academic department chairs, college deans, and university committees weighing research for promotion, tenure, or performance review consideration. The statement will be a pronouncement akin to the "Tenure, Promotion, and the Publically Engaged Academic Historian" report released by the American Historical Association, National Council on Public History, and the Organization of American Historians in 2010.
4. Luciano Floridi, *Information: A Very Short Introduction* (New York: Oxford University Press, 2010), 5.
5. One exabyte = 10^{18} bytes = 1 billion gigabytes = 50,000 year-long video of DVD quality.
6. Floridi, *Information*, 6.
7. Erez Aiden and Jean-Baptiste Michel, *Uncharted: Big Data as a Lens on Human Culture* (New York: Riverhead Books, 2013), 16.
8. "Wikipedia Community," *Wikipedia*, http://en.wikipedia.org/wiki/Wikipedia_community.

9. “Introducing Timeline,” *Facebook*, https://www.facebook.com/about/timeline.
10. John Palfrey and Urs Gasser, *Born Digital: Understanding the First Generation of Digital Natives* (New York: Basic Books, 2008), 40.
11. Ibid., 3.
12. Viktor Mayer-Schönberger and Kenneth Cukier, *Big Data: A Revolution that Will Transform How We Live, Work, and Think* (New York: Houghton Mifflin Harcourt, 2013), 6–7.
13. Aiden and Michel, *Uncharted*, 221, 236, 233.
14. Ibid., 23.
15. William Grimes, “Big Data Becomes a Mirror,” *The New York Times* (December 24, 2013), http://www.nytimes.com/2013/12/25/books/uncharted-by-erez-aiden-and-jean-baptiste-michel.html?_r=0.
16. “Principles and Best Practices,” *Oral History Association*, http://www.oralhistory.org/about/principles-and-practices/.
17. William Cronon, “How Long Will People Read History Books?” *Perspectives on History: Newsmagazine of the American Historical Association*, 50(7) (October 2012): 6.
18. S. Shunmuga Krishnan and Ramesh K. Sitaraman, “Video Stream Quality Impacts Viewer Behavior: Inferring Causality Using Quasi-Experimental Designs” in *The Proceedings of the 2012 Association for Computing Machinery Conference on Internet Measurement* (Boston, MA: Association for Computing Machinery, November 2012), 221.
19. Nicholas Carr, *The Shallows: What the Internet Is Doing to Our Brains* (New York: WW Norton & Company, 2010), 7.
20. Stephen Ramsay, “The Hermeneutics of Screwing Around; or What You Do with a Million Books,” Lecture, presented at the Playing with Technology in History, Niagara-on-the-Lake, Ontario, April 17, 2010, http://www.playingwithhistory.com, 3.
21. Siva Vaidhyanathan, *The Googlization of Everything (And Why We Should Worry)* (Berkeley: University of California Press, 2011), 201.
22. Bill Wasik, *And Then There's This: How Stories Live and Die in Viral Culture* (New York: Viking Penguin, 2009), 5.
23. Maria Konnikova, “The Six Things that Make Stories Go Viral Will Amaze, and Maybe Infuriate, You,” *The New Yorker* (January 21, 2014), http://www.newyorker.com/online/blogs/elements/2014/01/the-six-things-that-make-stories-go-viral-will-amaze-and-maybe-infuriate-you.html.
24. Eli Pariser, *The Filter Bubble: How the New Personalized Web Is Changing What We Read and How We Think* (New York: Penguin Books, 2011), 3.
25. Greg Bensinger, “Amazon Wants to Ship Your Package Before You Buy It,” *Wall Street Journal* (January 17, 2014), http://blogs.wsj.com/digits/2014/01/17/amazon-wants-to-ship-your-package-before-you-buy-it/tab/print/.
26. Pariser, *The Filter Bubble*, 101.
27. Mayer-Schönberger and Cukier, *Big Data*, 166.
28. Dave Eggers, *The Circle: A Novel* (New York: Vintage Books, 2013), 207.

CHAPTER 11

[o]ral [h]istory and the [d]igital [h]umanities

Dean Rehberger

The purpose of this last part of *Oral History and Digital Humanities* is to be a capstone, a stopping place that ponders the connections between the two areas of study. To do so, we should begin with a definition of the Digital Humanities (DH for short). Fortunately, digital humanists have a mania for defining the Digital Humanities, almost an obsessive compulsion. A bit of googling for variations of the phrase, "defining the digital humanities" turns up thousands of websites and hundreds of blog posts and articles, as well as a twitter stream @DefiningDH. In the last few years a spate of books have been published that attempt to delineate and define the field. Any or all would be a great starting point for those new to Digital Humanities or for those who desire a deeper understanding.

Matt Gold's collection of original articles and expanded blog posts in *Debates in the Digital Humanities* (2012) is a wonderful place to begin. Not only does it collect "together leading figures in the field to explore its theories, methods, and practices," but one can also find a free open-access edition online.[1] The collected works nicely outline the cracks and fissures currently vexing DH. Claire Warwick, Melissa Terras, and Julianne Nyhan's collection of essays, *Digital Humanities in Practice* (2012), takes us from the debates to *doing*. Based on ongoing research and teaching, the essays explore the various ways that digital humanists go about their projects, from their work, to their social media use, to their differing institutional contexts.[2] The collection ranges from crowdsourcing transcriptions to the computational reconstruction of ancient frescoes. Melissa Terras and Julianne Nyhan along with Edward Vanhoutte followed this up in 2013 with *Defining Digital Humanities: A Reader*, a wonderful collection of pivotal and sometimes controversial articles.[3] The collection includes such works as Mark Sample's "The Digital Humanities is not about Building, it's about Sharing," Stephen Ramsay's "Who's In and Who's Out" and "On Building," and Matthew Kirschenbaum' s "What Is Digital Humanities and What's It Doing in English?" In his *Understanding Digital Humanities* (2012), David Berry brilliantly renders the "computational turn" that is taking place in the Arts and Humanities and the ways this turn is both providing new avenues for research and new collaborative and interdisciplinary structures of research. All in all, the works map out a collaborative and vibrant field, yet one filled with tensions and uncertainties about its institutional place and future directions.

From the year 2012, however (2012 does seem to be the watershed year for books defining the digital humanities), the work that might prove most interesting to oral historians is *Digital_Humanities*, a book whose very production—written collaboratively by five authors, Anne Burdick, Johanna Drucker, Peter Lunenfeld, Todd Presner, and Jeffrey Schnapp—speaks volumes about the new ways digital humanists are going about their practice.[4] Not only does the work deal more with media, but one of the central arguments is that the digital age is taking us back/forward into a new age of orality. The aural as well as the visual are returning to the center stage. As they argue, "the Digital Humanities all contribute to the resurgence of voice, of gesture, of extemporaneous speaking, of embodied performances of argument."[5] This echoes directly Sherna Berger Gluck, Charles Hardy, Will Schneider, Tom Ikeda, and Gerald Zahavi, and their unwavering focus on the "aural," the importance of the embodied voice transmitted in context. This is not new for oral historians but they see the digital as a means for transmitting the full richness of the stories and voices. As Charles Hardy explains, "[a]ural history as lived only occupies the moment it is spoken. Once recorded, and now captured in space, it can be reanimated in new temporal moments and contexts." We will return to this thread later. For now, we must continue our quest to define DH.

In addition to the blogs, articles, books, and tweets—as part of the mania—each year we gather, virtually, for the Day of DH and post our personal definitions of the Digital Humanities as well as a blog about what we do in a typical day as digital humanists. This year, Day of DH 2014 attracted over 500 participants from around the globe. The definitions appear to be widely disparate and yet converge. A few examples should suffice.[6]

Aris Xanthos, University of Lausanne: "DH is a term that covers all scholarly methods, practices, endeavors, and challenges lying at the intersection of Humanities and Information science and technology."

Andie Silva, Wayne State University: "Any academic, archival, pedagogical, or project-based work that involves using, producing, and disseminating technology in the Humanities."

Caroline Sporleder, Trier University: "For me, Digital Humanities mean combining digital representation and novel automatic analysis techniques for textual and non-textual data in order to address new research questions from the Humanities and answer old questions better or differently. Ideally innovation in this process should happen both on the Humanities and the technology side."

Caroline McGee, Trinity College Dublin: "The organic nature and exciting potential of digital humanities makes defining the discipline a challenge even for experienced practitioners. With that in mind, could Oscar Wilde be right—'To define is to limit'?[7] Based on the work I do, DH is the application of digital technologies to traditional data and artifacts in order to make new interactive resources for those who are teaching and learning history."

Matthew Hiebert, Electronic Textual Cultures Lab: "The intersection of humanities inquiry and digital technologies."

Nate Sleeter, George Mason University: "It's the humanities, but it's also digital."

While the definitions all have a different valence, they all intersect at the different ways the humanities and technology converge for research and teaching. What is most interesting here is that one could replace the terms "digital humanities," "humanities," and "arts" with "oral history" in the definitions above and they could appear to fit into any of the preceding works in this book: "It's oral history, but it's also digital."

Along these lines, one of my favorite definitions is by Dan Cohen. Cohen is the former director of one of the largest digital humanities centers in the country, the Roy Rosenzweig Center for History and New Media at George Mason University,[8] and is now founding executive director of the Digital Public Library of America.[9] At the Day of DH 2011, he defined the Digital Humanities:

> Broadly construed, digital humanities is the use of digital media and technology to advance the full range of thought and practice in the humanities, from the creation of scholarly resources, to research on those resources, to the communication of results to colleagues and students.[10]

With a strategic replacement of words, Cohen's definition nicely encapsulates the works that appear in *Oral History and Digital Humanities:*

> Broadly construed, *digital* oral history is the use of digital media and technology to advance the full range of thought and practice in oral history, from the creation of scholarly resources, to research on those resources, to the communication of results to colleagues and students.

While this is a satisfying definition, it is missing one key element that is a thread in all of the works in *Oral History and Digital Humanities*: oral history, in its collaboration between researcher and narrator, is a generative and creative space that produces both something new and something more than the words that a transcript can capture.

We need to push the definition of DH further. Having worked on the Quilt Index project,[11] one of the largest online repositories of quilts, for a number of years, I find that a more fruitful way of approaching a definition of DH is through metaphor. I define DH in terms of quilt work. Quilting has seven defining elements that work for our purposes here:

1. Quilting can be done by an individual but often, traditionally, it is a community act of many hands, a collaborative enterprise.
2. It is about making and doing.
3. It is about reuse and remix.
4. It is both an art and a science.
5. It is generative of something new and transformative.
6. It is both public and private.
7. It has been historically devalued.

Of these elements, the first five are self-explanatory and easily map on to definitions of the Digital Humanities and Digital Oral History (DOH). While DH is about applying digital and network technologies to Humanities research and teaching, it is also about a way of doing work.

In the Humanities, the traditional model is that of the lone scholar toiling away in the archive or library only to emerge with a fully realized article or monograph. DH is about making things and not always successfully. Yet as in quilt making and sciences, failure is always acceptable and often instructive. One simply pulls out the threads and starts again. What is most wonderful about the early work on Digital Oral History is the drive to move forward despite the limits of the early media technologies and the delight in reporting mistakes as learning moments. As Douglas A. Boyd notes, when he realized he removed the natural hissing sound of a turf fire from a recording and found it a "very powerful lesson for [him] in terms of the use of and control of technology in recording and curating oral history."[12] Beyond a different sense of success and failure, sharing marks one of the cornerstones of the DH world. Much of what is done in DH, since often underfunded, is to reuse, remix, and share open-source code and projects. The key to this definition, however, is the double meaning of element five: "It is generative of something new and transformative." With collaborative making, not only do we emerge with a product, but the very process itself is transformative. When community group members gather to exchange expertise and build, whether quilt making or Quilt Index making, the process transforms those who participate. It can't help but do so.

While the first five elements of my definition allow me to repeat core features of the DH community, element six might stop us up. It does not appear to map so easily. Quilts are both public and private objects since they are often displayed but have utilitarian use, keeping us warm. This is, in a similar but slightly altered fashion, true of most DH and DOH projects. Often the results of the projects are to be used by the researchers for their work and studies. Yet when made accessible in online repositories, digital exhibits, and applications, they often have broad public appeal. This is true, for example, of the Quilt Index and its associated oral history project, Quilters' S.O.S.—Save Our Stories.[13] On the one hand, the sites are excellent resources for material culture and quilt scholars. On the other hand, they also have wide appeal among quilt makers and collectors, the general public, and surprisingly (from the analysis of user logs), a sizable number of users coming from the military.[14]

Perhaps the seventh element is not so strange for many of us. For many areas of study dominated by women practitioners, there has been a long and ongoing battle for legitimacy. For a long time (and still by some), quilts were devalued as simply craft: unimportant work done predominately by women. Now many recognize quilts as significant, historically valuable, cultural products and artworks. For those who work in Digital Humanities, Oral History, Public History, there has been a similar battle for legitimacy. DH is often devalued as service work or lacking rigor. Unlike the measurable output of peer reviewed articles and books in the traditional disciplines of the Humanities, DH products appear strange and difficult to count. For Oral History, as a field, the legitimacy questions are even more complex and tendentious and do not need to be rehearsed here. Suffice it to say here that underlying many of the works in *Oral History and Digital Humanities*, we see pioneering work that both struggles with the early technologies—slow, buggy equipment and software—and struggles to legitimize the recording over the transcript, the voice over the document.

Although we have a ways to go to the conclusion, the conclusion I am working toward is rather obvious. Digital Oral History is an allied field of the Digital Humanities. Indeed, it is perhaps the most fully realized allied field of DH. In the end, this book is talking about oral history projects and telling us their digital story. In an obvious way, all Digital Oral History projects are Digital Humanities projects because Oral History is part of the Humanities. However, DH and DOH share a much deeper bond. They, indeed, share the same challenging spaces inside and outside the academy:

1. DH and DOH methods find homes in both the Humanities and the Social Sciences.
2. Work in DH and DOH is found in a host of disciplines and area studies: history, anthropology, folklore, English, rhetoric, American studies, cultural

studies, women's studies, Chicana/o studies, African American studies, Jewish studies, media and film studies, and more.

3. DH and DOH projects appear in libraries, archives, and museums, public and private, as well as other governmental and non-governmental agencies (NGOs).
4. DH and DOH projects find labor in many of the same places for programming, transcription, and design: student labor, non-profits, and for-profit vendors.
5. DH and DOH both profit and thrive by having centers. The many Oral History and Digital Humanities centers around the world not only help to establish and support local projects but the larger and more long-standing centers help to establish, distribute, and teach methods and best practices.
6. DH and DOH both struggle with the big tent. In ways, Oral History has always struggled with a much bigger tent. Like DH, it has constituency across the disciplines and outside the academy in both the public and private sector, but Oral History also has a larger constituency of citizen practitioners and storytellers.
7. Because of the big tent struggle in both DH and DOH, there are those who call for more strict methods and a highly defined field while others call for more openness and sharing.
8. And finally, on a most basic level: In the "Introduction," Doug Boyd and Mary Larson point out "[i]t was technology from which oral history was born." The same, of course, is true of the Digital Humanities.

DH and DOH, beyond sharing values, methods, and ways of working, also share structural and organizational tensions and strains.

Along these lines of sharing, we can turn to one last, and most important, point of convergence of DOH and DH. Similar to the ways in which Oral History shifted with the pressure of Social Historians to capture the voices of ordinary citizens and those often excluded and silenced by history, an important strand of DH came from the same impulses. Roy Rosenzweig, a labor and public historian, founded the Center for History and New Media (CHNM) at George Mason (now renamed the Roy Rosenzweig Center for History and New Media). Rosenzweig was an activist and much of his digital work focused on making the struggles and everyday lives of ordinary Americans—workers, ethnic minorities, women—visible and comprehensible to a wide audience, and he believed in the democratizing power of technology.[15] Orville Vernon Burton was the founding director of the Institute for Computing in Humanities, Arts, and Social Science (I-CHASS) at the University of Illinois and is a social historian and noted historian of slavery. Mark Kornbluh founded Matrix at Michigan State University and again is a social historian and has done extensive international DH outreach and collaboration, particularly in Africa. The social historians Ed Ayers and Will Thomas founded the Virginia Center for Digital History (VCDH) at the University of Virginia and worked with Institute for Advanced Technology in the Humanities (IATH) at the University of Virginia to do the a milestone digital humanities project, *Valley of the Shadow*. While a lot of the work of these centers was to produce resources for scholars and students, all had the underlying impulse that the public work of the digital humanities does important social and cultural work.

While the impulse for Sherna Berger Gluck, Charles Hardy, Alessandro Portelli, Will Schneider, Tom Ikeda, Doug Boyd, and Gerald Zahavi is to capture and reproduce the fullness and richness of the oral history experience, they also do what they do because of the firm belief in oral history as an agent of social change. As Tom Ikeda explains,

> This question shifted the discussion from simply conducting and preserving oral history interviews to a much broader discussion of how the stories could be used to promote and advance the country's democracy... Densho decided that it would be important to design a project that did more than preserve the Japanese American incarceration story; the project also needed to share the stories in a way that kept the stories alive so that similar targeting would not happen to another group.

Digital media and network technologies afford researchers the ability to reach out to a wide audience. Charles Hardy wonderfully notes this reach in his essay after he raps out a long list of possible audience members that includes all "who engaged the beauties and mysteries of life through their ears." But it is only effective because digital media and network technologies allow for a more fully contextual experience that goes beyond the flat words on a page and takes the listener not only into the experience of the actual speaker's voice, gestures, and tone but allows the researcher to surround the voice with all the other needed cultural and historical cues to help the listener/viewer understand the oral history: for example, maps, documents, customs, folklore, histories. At the very least, Will Schneider argues, we need to capture and supply all details when "chronicling the circumstances of the recording, the intent and interest of both recorder and teller, previous recountings of the information on the part of the speaker, and some historical and cultural context on the subjects discussed." Only by doing so can we fully try to re-present the complexity of the narrative moment.

At the heart of the matter—and what makes this all work for both DOH and DH—is metadata. After all the talk of generative discourse, transformative practice, and social change, the engine at the bottom of it all is the data about the object. We not only need to capture the full context of the object produced—oral history, image, video—but we need to do it in a regular and meaningful way so it can be processed both by machines and humans. This is precisely the point

Elinor Mazé makes. Metadata not only makes objects searchable and useful in applications but enriches the experience of the object. As Mazé explains the work of the archivist, "[w]e strive now also to enrich that collection, to broaden and deepen the understanding of each of its stories, to reveal many facets of their significance to diverse populations, and to do all of this in ways that are somehow likely to survive rapid, never-ending changes in all of the tools of preservation, access, and analysis." In my darker hours, when I doubt the transformational and social impact of DH and DOH, I do come to believe that if nothing else, DH and DOH teach scholars about the importance of structured data. This was, in fact, one of the primary conclusions of our Oral History in the Digital Age project.[16] Oral Historians are the first archivists on the scene and must capture a robust set of data to contextualize the oral history. But that is for another essay.

As we move toward conclusions, we need to turn to the *lowercase* digital humanities. As we have seen, the *uppercase* Digital Humanities refers to an institutional affiliation, a way for people in universities, libraries, archives, and museums to identify with a field and research identity. The lowercase digital humanities refer to an age, an historical period, in which much of the humanities we produce and consume is in digital form. There is no reason here for me to re-explain the digital tsunami we are experiencing. Stephen Sloan does a masterful job of illustrating how quickly the developed world has become digital:

> In many respects, our society went from analog to digital at the speed of light. Just over a decade ago, only a quarter of the information stored globally was digital. A significant majority of recorded information was held in analog formats, including paper, film, tape, and vinyl. Since that time, however, the proportion of analog to digital quickly reversed. By 2013, the explosion of digitally stored information meant that now less than 2 percent is nondigital. In the span of just a few decades, all members of the G7 group—Canada, France, Germany, Italy, Japan, the United Kingdom, and the United States—were transformed into information societies where now at least 70 percent of gross domestic product relies on intangible goods (information related) rather than material goods (physical agricultural or manufacturing production).

The humanities outside academia in developed nations is digital. People consume and produce primarily digital media. Perhaps what shows the thriving nature of the lowercase oral history and the digital humanities is to do a search for the term, "oral history" on YouTube.[17] While the occasional zombie and spoof oral history comes up, many of the search results are an excellent cross-section of oral history materials and projects, a wide range of materials created inside and outside of the academy. At the same time, the search results show the amazing diversity and vitality of the field, from labor history to LBGT projects, from methods to recording tips. The same is true if similar searches are done on Vimeo[18] and

SoundCloud.[19] These search engines become a space where Oral History converges with oral history; Digital Humanities converges with digital humanities.

At this point, with many conflicting studies, we are not very sure what impact living and working in the digital humanities will have on our students and us. Some claim that it shortens attention spans and Google is dumbing down society. I do know that students can spend countless hours weaving through highly complex narratives of video games. What this means I am not sure. Much of the evidence we have is anecdotal or found in studies that have not been replicated. We must remember that Plato in his dialogue, *Phaedrus*, saw writing as a kind of "poison" that would greatly hurt memory. And writing did indeed have a great and uneven impact on oral cultures that can be viewed in both positive and negative lights. All new information technologies have an impact, however uneven and contested: the printing press, mass circulation libraries, radio, television. Often it takes distance to see this impact.

It is precisely in this space that we can use oral history as disruptive technology. As Marjorie Mclellan notes in her essay, oral history is an exceptionally active and experiential learning exercise. While reviewing oral methods, I have students track their own digital media and social networking use. Then I have students interview a parent or a grandparent about how they maintained reputation and socially interacted with other students in high school before the Internet and the proliferation of digital technologies. Students love the assignment and learn about all kinds of wonders like party lines, princess phones, Walkmen, beepers, and movie theater balconies. As always, many report liking the interview process and walking away with a new sense of their family member. The class always ends up nicely divided between the past and present and makes for lively discussion about the positives and negatives of past and present socializing. That is, as we move forward, it may be Oral History projects that help us to understand the digital humanities. Projects that are much more formal than a classroom project, but classroom projects do help.

Returning to the uppercase Digital Oral History and Digital Humanities. The questions remain: does it really matter? Why identify with one field or the other? Wouldn't it be better to focus energies on developing the field of Digital Oral History rather than being allied with an ill-defined field like the Digital Humanities that seems to be going through its own struggles and growing pains? There are no simple answers to these questions other than that many of us have multiple identities within institutions: administrator, historian, folklorist, oral historian, digital humanist. The value of the identity is to share space with others who similarly identify. Institutional identities allow us to gather at meetings and conferences and share work on websites and blogs, with journals and books.

In his "Orality" piece, for example, Doug Boyd describes his extremely valuable Oral History Metadata Synchronizer (OHMS) project. Is this an archive and libraries project? An oral history project? Or a Digital Humanities project? The answer, of course, is all three. Feedback and exchange from all three communities was instrumental in developing the project. More important, though, the larger user base and community allow for more creative and adaptive thinking. Doug has a wonderful eureka moment in his piece when he explains,

> I recall that as a researcher, I love the granular search provided by the transcript. As an administrator, I love the budgetary opportunities created by indexing. However, users and researchers engage with oral history in different ways. I have come to the conclusion that the better user experience is when the OHMS viewer presents both an OHMS Index and a transcript, allowing the user to toggle between both discovery tools as needed. But the archive no longer has to hold interviews hostage on the physical or virtual shelves for decades, awaiting funding for transcription.

OHMS now has value for not only Oral Historians and Folklorists but also for anyone who wants to organize, tag, index, and search media: not for only personal archiving but for anyone in the digital humanities and social sciences who has to work with media interviews and footage.

In addition to the community and resource benefits of institutional identities, they also offer strategic benefits. It is often very difficult for individuals to get funding for a project, but creating research groups that are interdisciplinary and inter-institutional opens up more funding opportunities. And finally, identifying with the Humanities portion of the Digital Humanities has political importance. Politicians often do not have fine-grained hammers when it comes to the Humanities. They will rarely say, "Let's cut the Humanities but save those cute little oral historians in the corner." Saving and transforming the Humanities is part of all of our identities, and particularly for those who want to save the human voice and story.

Notes

1. *Debates in the Digital Humanities*, ed. Matthew Gold (Minneapolis: University of Minnesota Press, 2012) can be found at http://dhdebates.gc.cuny.edu/.
2. *Digital Humanities in Practice*, ed. Claire Warwick, Melissa Terras, Julianne Nyhan (London: Facet Publishing in association with UCL Centre for Digital Humanities, 2012).
3. *Defining Digital Humanities: A Reader*, ed. Melissa Terras, Julianne Nyhan, and Edward Vanhoute (Farnham: Ashgate, 2013).
4. *Digital_Humanities*, ed. Anne Burdick, Johanna Drucker, Peter Lunenfeld, Todd Presner, and Jeffrey Schnapp (Cambridge: MIT Press, 2012).
5. *Digital_Humanities*, Kindle Edition (2012), 11.

6. All the following quotes were taken from the Day of DH 2014 website, accessed May 5, 2014, http://dayofdh2014.matrix.msu.edu/.
7. Oscar Wilde, *The Picture of Dorian Gray*, (1891) (CELT: Corpus of Electronic Texts), accessed April 3, 2014, 217.
8. Information on the Roy Rosenzweig Center for New Media can be found at http://chnm.gmu.edu/.
9. Information on the Digital Public Library of America can be found at http://dp.la/.
10. Dan Cohen, "Defining Digital Humanities, Briefly," *Personal Blog*, accessed May 5, 2014, http://www.dancohen.org/2011/03/09/defining-digital-humanities-briefly/.
11. Information on the Quilt Index can be found at http://www.quiltindex.org/.
12. Doug Boyd, "Case Study: Noise Reduction and Restoration for Oral History–The Stars of Ballymenone," in *Oral History in the Digital Age*, ed. Doug Boyd, Steve Cohen, Brad Rakerd, and Dean Rehberger (Washington, DC: Institute of Museum and Library Services, 2012).
13. Information about the oral history project Quilters Save Our Stories can be found at http://www.allianceforamericanquilts.org/qsos/.
14. We have not done follow-up user studies to find out why this is, but perhaps they offer comfort when far from home or there is a sizable number of quilt makers in the military.
15. Biographical information taken from the Roy Rosenzweig memorial site at http://thanksroy.org/about, accessed May 7, 2014.
16. http://ohda.matrix.msu.edu/.
17. https://www.youtube.com/results?search_query=oral+history.
18. http://vimeo.com/search?q=oral+history.
19. https://soundcloud.com/search?q=oral%20history.

Index

CPI Antony Rowe
Eastbourne, UK
September 04, 2019